RENEWABLE ENERGY

MADE EASY

Free Energy from Solar, Wind, Hydropower, and Other Alternative Energy Sources

By David Craddock

RENEWABLE ENERGY MADE EASY: FREE ENERGY FROM SOLAR, WIND, HYDROPOWER, AND OTHER ALTERNATIVE ENERGY SOURCES

ISBN-13: 978-1-60138-240-5   ISBN-10: 1-60138-240-5

Library of Congress Cataloging-in-Publication Data

Craddock, David, 1982-
  Renewable energy made easy : free energy from solar, wind, hydropower, and other alternative energy sources / by David Craddock.
      p. cm.
  Includes bibliographical references and index.
  ISBN-13: 978-1-60138-240-5 (alk. paper)
  ISBN-10: 1-60138-240-5 (alk. paper)
  1. Renewable energy sources. 2. Renewable energy sources--Economic aspects. 3. Energy policy. I. Title.

  TJ808.C73 2008
  621.042--dc22
                          2008024787

Printed on Recycled Paper

INTERIOR LAYOUT DESIGN: Nicole Deck ndeck@atlantic-pub.com

Printed in the United States

Author Acknowledgements

Many established authors are lucky enough to expend as little as two words in their acknowledgments section. "To Mary," or "Hey Fred," or "You know who you are" might work for those folks, but for this first time author, "Howdy" just will not cut it. So without further ado, this book is dedicated to:

Angela Adams and the editors of Atlantic Publishing. Thank you for helping me make this book the best it can be.

Teachers, professors, and authors that never failed to inspire: Robert King, Terry Sosnowski, James Christine, Terry Goodkind, R.A. Salvatore, and Greg Keyes.

Mom, for only allowing me to play Nintendo 30 minutes a day, which gave me more time to read. Debbie and Dave, for never failing to bellow "WEE!" when I need to hear it most. Rita: I hope this earns me a spot in your family tree. Grandpa Bob, for always telling me to "Hang in there, buddy." Grandma Shirley, for telling me to keep writing and for showing off my work to anyone within a five-mile radius.

Holly, Binx, Yuki, and Onyx — but not Bucket.

Uncle Brad, Aunt Cindy, Shawn, and Julia. Thank you for all of the help since my big move.

My best friends: Daisy, Andrew, Jeff, and "Common Red Knowledge."

Mom-Teresa: I love you "muchly!" Jennifer and Tony, some "pretty cool guys" (with one obviously being a female, but still "pretty cool" nonetheless). Daniel, my brother: go buy this book, jump up on your car, and do The Rock's pose for me!

Grandma and Grandpa Craddock, for always showing support and encouraging me to take lead during my cart-pushing meetings. That parking lot was always messy — but more so when I was there.

Dad. "Son, act like you've got some sense." Never!

My ever-present muse. Thank you.

And last but never least — Amie Christine, my love, my life. Thank you for putting up with me even when I don't want to.

We recently lost our beloved pet "Bear," who was not only our best and dearest friend but also the "Vice President of Sunshine" here at Atlantic Publishing. He did not receive a salary but worked tirelessly 24 hours a day to please his parents. Bear was a rescue dog that turned around and showered myself, my wife Sherri, his grandparents Jean, Bob and Nancy and every person and animal he met (maybe not rabbits) with friendship and love. He made a lot of people smile every day.

We wanted you to know that a portion of the profits of this book will be donated to The Humane Society of the United States.

–Douglas & Sherri Brown

THE HUMANE SOCIETY
OF THE UNITED STATES ©

The human-animal bond is as old as human history. We cherish our animal companions for their unconditional affection and acceptance. We feel a thrill when we glimpse wild creatures in their natural habitat or in our own backyard.

Unfortunately, the human-animal bond has at times been weakened. Humans have exploited some animal species to the point of extinction.

The Humane Society of the United States makes a difference in the lives of animals here at home and worldwide. The HSUS is dedicated to creating a world where our relationship with animals is guided by compassion. We seek a truly humane society in which animals are respected for their intrinsic value, and where the human-animal bond is strong.

Want to help animals? We have plenty of suggestions. Adopt a pet from a local shelter, join The Humane Society and be a part of our work to help companion animals and wildlife. You will be funding our educational, legislative, investigative and outreach projects in the U.S. and across the globe.

Or perhaps you'd like to make a memorial donation in honor of a pet, friend or relative? You can through our Kindred Spirits program. And if you'd like to contribute in a more structured way, our Planned Giving Office has suggestions about estate planning, annuities, and even gifts of stock that avoid capital gains taxes.

Maybe you have land that you would like to preserve as a lasting habitat for wildlife. Our Wildlife Land Trust can help you. Perhaps the land you want to share is a backyard—that's enough. Our Urban Wildlife Sanctuary Program will show you how to create a habitat for your wild neighbors.

So you see, it's easy to help animals. And The HSUS is here to help.

The Humane Society of the United States
2100 L Street NW
Washington, DC 20037
202-452-1100
www.hsus.org

TABLE OF CONTENTS

CHAPTER 3: POWERING MR. FUSION - BIOMASS AS AN ENERGY SOURCE   51

CHAPTER 4: GETTING WET AND WILD WITH HYDROPOWER 69

CHAPTER 5: RIDING THE WINDS OF CHANGE - WIND POWER 85

FOREWORD

The world is at a crossroad. Down one path lie dirty skies, rising oceans, melting glaciers, and the suffering of billions. Down another we have a clean future where people live in a green and sustainable way. Look around you; the people currently inhabiting this planet are in a unique position. We have been tapped to choose the future of all humanity. It is a big responsibility; the choices that we make today will affect the future of the human race. I hope we choose wisely.

A vital component of that green sustainable future is choosing to feed humanity's energy needs from sources other than fossil fuels. For too long we have had a free ride, blissfully burning carbon rich fuel sources to power our lives. The miracle of coal, oil, and gas came with a hidden price, global warming. Over the last 30 years, during an ever growing level of concern about our carbon rich lifestyles, some of the world's most intelligent people have been working hard to replace fossil fuels with something much cleaner and sustainable.

Renewable energy comes from sources that are easily replenished. Sources such as solar, wind, geothermal, wave, and tidal power offer a promise of a future free from the dangers of global warming, wars for oil, and pollution. There is a race going on. Instead of trying to reach the moon, scientists, back yard inventors, students, and every day folk have been hard at work building ways to harness the energy sources of the future.

Renewable Energy Made Easy shows how renewable energy

is changing the world, and that you too can participate in the revolution. Retrofit your house, build your own solar panel, erect a wind turbine, or support a wind farm. Whatever you decide to do, this book will show you that it is not hard to make the world a better place.

Renewable energy is the fastest growing source of energy on this planet. People are hard at work putting up wind turbines, installing solar panels, and are building a whole host of clean sustainable sources of energy. As the cost of fossil fuels grows ever higher, renewable energy looks more and more like the affordable way to power our life. It is increasingly clear that renewable energy is not only the sustainable way to produce power, but also the most economical.

Every day the wind blows, the sun shines, the waves crash, and the earth turns. These energy sources are out there waiting for us to capture them and put them to good use. Perhaps in 100 years the oldest of us will talk about the end of the fossil fuel obsession and how our generation took up the challenge and led humanity down the path towards a renewable energy future.

~ *Shane Jordan, Environmental Campaigner and Director of Education and Outreach for the Massachusetts Bicycle Coalition—a statewide bicycle advocacy group*

Credentials: Computer science degree, general science background, several years working in the renewable energy field mostly in home scale renewable energy retrofits and new construction, extensive experience in environmental restoration with a focus in dry land forest and coastal grasslands, head editor **TheSietch.org** and lead author for The Sietch Blog. Inventor, citizen scientist, environmental campaigner, and bicycle advocate.

Bio: Shane Jordan has taken a long and strange path towards making the world a better place. What started as an obsession with computers has led him to take an ever more diverse and green series of jobs. He has repaired trails in high altitude desert, been part of grassland restoration projects in the marshes of Cape Cod, installed wind turbines, taught kids how to build solar powered models cars, and helped design Web sites for environmental campaigners. He currently finds himself enjoying the thrill of riding his bicycle over 100 miles a week on the streets of Boston, while working to promote the bicycle as an active (and green) form of transportation for the citizens of Massachusetts. When he isn't running **TheSietch.org** he likes to invent renewable energy experiments that people can make themselves. His newest project is a website for bicycle enthusiasts in Boston called **BostonBiker. org**.

TheSietch.org

REDUCE, REUSE, AND RENEW

1.1: THE ALMOST-PERFECT DAY

Sixty-five degrees Fahrenheit. No clouds in the sky. A light breeze that ruffles your shirtsleeves and teases your skin.

Your keys beckon from the edge of your desk, and without a second thought, you scoop them up and bound out of the house.

The procedure is familiar and fun: hop in the car, roll down the windows, turn the key, and shift into drive, causing the engine to sputter for a moment before it roars into happy life and settles into a purr. You press down on the gas and begin to glide forward, laughing as the wind runs through your hair. Where are you going? It does not really matter. You are the recipient of a perfect day with endless possibilities.

Suddenly, the engine begins to cough and sputter. Concerned, your eyes flicker to the gas gauge where the red needle is wavering near 'E'. You smile. This is no problem. All you have to do is pull over into the nearest gas station, feed a $5 bill to the attendant, and you will be back on the road with a full tank.

Yeah, right. When was the last time $5 bought you a full tank of gas? For some readers, the answer could be "Never." It is a tricky game the government is playing with our gas prices. A few years back, around 2001 or so, unleaded gasoline sold for around $2.50 a gallon. Fast forward to the present, where prices in most areas are around $3.20 a gallon. While this may make you miss the days when $2.50 seemed expensive, $2.50 is not cheap; it is only economical when compared to modern-day prices.

Gas prices are rising higher, and they show no signs of evening out anytime soon. Even if they did even out, would we accept $3.20 per gallon as "cheap" if we could keep the price from rising to $4, $5, or worse? Gasoline is becoming scarce, along with many other fossil fuels, so why not take advantage of materials that are a part of our everyday lives? Animal waste, air, sunlight, water, lumber, and ocean waves are all means to an end that could help save our children from a world where gasoline and other valuable fuels are beyond reach.

1.2: PROBLEMS WITH FOSSIL FUELS

Our dependency on fossil fuels is a large part of the problems we are experiencing. Fuels such as oil and coal power account for greater than 80 percent of worldwide consumption. Such widespread use is taking a toll on our planet, which means that if it is not already, it will soon be taking a toll on us.

Many of you probably know people who burn trash on their property. Though they are not exempt from reproach, these individuals account for only a small part of the overall problem. When large plants and factories go about burning mounds and mounds of trash, things get smoggy, stinky, and disgusting in a hurry. Burning fossil fuels results in the emission of harmful chemicals, such as various sulfur and nitrous oxides. One serious side effect caused by these emissions is the irregular changes in climate our planet has been experiencing for several years. Winters are shorter and summers are longer and hotter, and that is just the first of changes that we have noticed.

Harmful emissions are also responsible for countless health problems. If smoke and smog are devastating to our climate, you can imagine what it does to our bodies.

Even if the harmful side effects of using fossil fuels were to be eliminated, the problem of distribution would still exist. Many areas simply have more of some fossil fuels than others, which leads to unfriendly disputes, monopolies, and unfair barter agreements. For example, the Middle East is such a hotbed for oil raids because two-thirds of the world's reserves are stored there.

Just as bad as their harmful emissions and locations, fossil fuels are alarmingly finite. We have been using fossil fuels for centuries, but eventually, they will run out, which means suitable replacements need to be found posthaste. This is not a question of if fossil fuels will run out; it is a question of when they will be depleted.

1.3: RENEWABLE ENERGY - PROS AND CONS

Luckily, suitable replacements have been found in the

form of solar, hydro, wind, biomass, and geothermal energy schemes. These sustainable energy sources can be found almost everywhere, unlike the distribution concerns plaguing most fossil fuels. Other countries do not have to invade the Middle East to restock sunlight; all they have to do is look outside for this plentiful resource. Therefore, technologies such as photovoltaics, which rely on solar energy, can and will function anywhere sunlight exists.

Because a viable substitution for fossil fuels is becoming increasingly necessary, renewable energy is seeing an overwhelming abundance of support. While public opinion is quite high, many politicians are hesitant to embrace renewable energy due to cost, which is on average significantly higher than fossil fuels. Many costs related to renewable energy have dropped in recent years, but just like our example of gas prices, just because something cost less two decades ago does not mean it was "cheap" then, and it certainly is not considered "cheap" now. Some forms of renewable energy, such as geothermal, are somewhat competitive with fossil fuels, where as photovoltaic technology is expensive, and has been for quite some time.

Intermittency is another problem related to renewable energy. Some areas of the world, such as the Great Plains States in North America, offer vast wind resources; other areas do not. Put simply, if the wind is not blowing in a certain place, then wind power is not a viable solution for that area. Solar energy is a popular form of renewable energy, but what happens at night when the sun sets? An alternative form of energy — perhaps a fossil fuel — must be used. As you will come to see, hydropower is extremely effective, but can be extremely devastating if proper precautions are not taken to secure it.

1.4: THE FOCUS OF RENEWABLE ENERGY MADE EASY

Some forms of renewable energy carry more negative points than others. After making such points, one might wonder why we would pursue such forms of energy if, like wind power, they are only effective in certain areas. But when they work, they work well. Renewable Energy Made Easy will take you through the pros and cons of solar, biomass, hydro, wind, and geothermal energies. You will learn how these different sources work, why they work, what makes them efficient, and where they need to improve.

Just like fossil fuels, no substitute for the powerful trio of natural gas, oil, and coal will be perfect, but perfection is an unachievable goal. Science deals in facts, and it is a fact that a viable substitute to fossil fuels must be found and — more importantly — implemented as soon as possible. Wind, water, and sunlight will always exist; natural gas will not.

LET THERE BE LIGHT - THE POTENTIAL OF SOLAR POWER

2.1: INTRODUCTION

The sun is an ever-present force, rising every day and setting every night, just as it has since the world began.

In recent times, though, the gifts of sunlight have been overlooked. So reliant are we on our new-fangled electricity, furnace systems, and other technological gizmos that it is easy to forget how human beings survived before modern-day conveniences arrived. Even though electricity and a variety of other advancements have improved most of our lives, sunshine is still a viable solution for a wide array of problems.

2.2: A HISTORY OF SOLAR ENERGY

At some point or another, we have all encountered a playground bully, one of those punks who seemed perfectly content to bloody our noses, steal our lunch money, and participate in cruel acts, such as directing sunlight through a magnifying glass with the intention of lighting ants on fire.

Given the relative cruelty of the human race, it is not surprising that such a tactic is one of the first ways we made use of the of the sun — using a magnifying glass to set ants and grass aflame dates all the way back to seventh century B.C. During the ninth century B.C., many Greeks and Romans saw fit to apply a similar technique, using mirrors to light torches for religious ceremonies.

Perhaps one of the most popular myths ever told is that of Archimedes using bronze shields to direct sunlight at wooden ships used by the Roman Empire in second century A.D. As the legend goes, the "Death Ray's" magnification of the sun was so great that the Romans' wooden ships burst into flame and sank, thereby saving Syracuse, an Italian city on the eastern coast of Sicily, from its besiegement.

So popular was the "Death Ray" myth that a Massachusetts Institute of Technology group conducted an experiment on MythBusters, a popular television program airing on the Discovery Channel, which entailed setting fire to a wooden fishing boat in San Francisco. The group built a mirror comprised of 127 one-foot tiles, but they were unable to damage the ship beyond slight charring. When the show aired in January 2006, MythBusters declared the myth "busted."

More than three decades prior to the MythBusters episode, a Greek scientist by the name of Ioannis Sakkas used 70 mirrors coated with copper to set fire to a wooden mock-up of a Roman ship within seconds. A bit of controversy still surrounds Sakkas' experiment, as the ship was coated with tar, which is flammable and could have led to quicker combustion.

Solar energy has continued to grow in leaps and bounds over the past 2000 years. From the first through the fourth centuries, many Roman bathhouses made use of south-facing windows to capture sunlight and warm the structures, making for a relaxing environment. During the 13th century, Anasazi, descendents of Pueblos, lived in south-facing buildings that also captured the sun's warmth.

Progress continued in slow but steady bursts until 1839, when a French scientist by the name of Alexandre-Edmond Becquerel discovered the photoelectric effect, which dictates that sunlight can be converted into electrical energy when placed in certain cells. Shortly after, during the 1860s, French mathematician August Mouchet had the idea for solar-powered steam engines. With the help of his assistant, Abel Pifre, Mouchet worked tirelessly over the ensuing two decades before producing the world's first solar-powered engines.

Ensuing years saw the development of marvels such as solar cells made from selenium wafers, by Charles Fritts; the first solar powered water heater, by Clarence Kemp; Albert Einstein's Nobel Prize-winning theories on the photoelectric effect; the first telecommunications satellite, from Bell Telephone Laboratories; photovoltaic-powered residences; the flight of the first solar-powered aircraft; the Solar Challenger, by Paul MacCready; and much, much more.

2.3: WHAT IS SOLAR ENERGY?

"Solar energy," a term interchangeably used with "solar power," is simply energy from the sun. The sun has been, and can still be, used for many things: warming a house, heating a pool, providing light, burning ants, and sinking wooden ships — just to name a few.

But how, specifically, does the sun's energy aid us in even the most basic of human tasks, and what exactly are we feeling when we stand out in the sun during summertime?

We earthlings feel not even a fraction of the energy that the sun is capable of producing. The process begins in the upper atmosphere of the earth, which receives approximately 174 petawatts, or a whopping 1015 watts, of incoming radiation from the sun. This energy is being transmitted at any given time. Conditions such as weather, pollutants, and the atmosphere result in 89 petawatts actually being absorbed by earth's land and water sources.

The energy finally absorbed by the earth combines with the greenhouse effect — infrared radiation emitted by the atmosphere to warm the earth. The process of photosynthesis, through which living organisms such as plants convert light energy into chemical energy, helps to provide fossil fuels, wood, and even our food.

Because of constantly varying conditions, it is vital to have a thorough understanding of a climate in which the harvesting of solar power in anticipated.

2.4: WINDOWS (NOT THE OPERATING SYSTEM)

As powerful as the sun certainly is on its own, even schoolyard bullies understand that with something to magnify the sun's rays, extremely potent heat can be gained for various applications. A good "device" for collecting the sun's energies is a sheet of glass.

2.4.1: A History of Windows

The origins of the word "window" date all the way back to Old Norse, a Germanic language used by inhabitants of Scandinavia during the Viking Age. In Old Norse, the word for window, vinduaga, is a hybrid of the words vindr, "wind," and auga, "eye." During the early 1300s, Old Norse gradually faded into disuse, but their words were absorbed and adapted by other languages.

The ability to make glazed windows can be traced back to the Romans, who are credited with a procedure to make different plate glass, which, as we learned in Section 2.2, "they fitted into their bathhouses to provide warmth and comfort to those within.

Although the Romans were most impressed with the transparent properties inherent in glass, they often failed to make mention of another positive trait — its ability to withstand extremely high winds. Without that property, wind would have blown all through the bathhouses and other structures, which would have completely negated the light's warmth.

2.4.2: Windows in Action

Glass is a magnificent magnifier of sunlight, largely due to certain properties of transparency that it possesses. Glass is completely transparent to both short-wave infrared radiation and visible light. At the opposite end of the spectrum, glass is impervious to long-wave infrared re-radiated from a solar collector (discussed in section 2.5, "Solar Collectors"), or any building behind a solar collector.

Like any area of technology, the past several decades have seen enormous effort spent trying to improve the glazing on windows. This research has two primary goals: to improve glazing's transparency to the sun's visible rays, and to thwart heat from escaping through the window.

It is the goal of window manufacturers to make every window as transparent as possible; this is known as maximizing the window's transmittance, which is the fraction of incident light that passes directly through the glass. Manufacturers typically try to maximize transmittance by making the iron content in pieces of glass as low as possible.

Plastics, also prized for their transparency, often contain optical properties very similar to those found in glass, and therefore can sometimes be used as a substitute. However, they have one major failing when compared to glass, which is that plastics must be outfitted with some sort of protection against ultraviolet light.

2.4.3: Heat Loss

As mentioned, much research and expenditure has gone

into making sure windows do not give up more heat than they gain, or even more preferably, any heat at all. Even when shut tight and properly sealed, heat always manages to escape. Largely, the reason stems from a peculiar trait of heat. Heat will pass through any given substance if the heat on either side of the substance is different. How often heat will pass through and how much heat will be lost can be determined by considering a few key factors.

- The difference of temperature A and temperature B, with each temperature variable being on its own side of a piece of glass.

- The amount and quality of insulation material used on the glass.

- The maximum amount of the glass available for proper heat flow; the larger a window, the greater potential for heat loss.

Heat loss cannot be entirely eliminated, so it would be more practical for us to focus on learning how to minimize heat loss. To do this, we must first come to understand what is entailed in transmitting heat from place to place.

In any given substance, heat will be transferred back and forth between hot and cold regions. This process is called conduction. To determine the rate of conduction's flow, it is important to know the difference of heat between each side of the substance — temperature A and temperature B — as well as the maximum amount of glass available for the heat's flow.

Also integral to determining the rate of conduction's flow

is the thermal conductivity of the substance, defined as a measure of how well a material can transfer heat. Metals are the material known to have one of the highest rates of thermal conductivity, and they are able to swap enormous amounts of heat back and forth with small temperature differences.

Insulators increase the difference in temperatures needed to transfer heat, and still air is a great insulator. One effective method of properly insulating panes of glazing is to put pockets of air in between the glass panes. Air bubbles can be placed in plastic material, and even between fibers of wood.

After a fluid or gas has been warmed, it will begin to expand and decrease in density as it begins to rise. Such a process is called convection. Convection is one of the most common ways that air leaves windows and makes its way outside.

The process of convection occurs between the glass and air stationed on the exterior and interior surfaces. That is what happens when dealing with single-glazed windows; double glazing's convection process focuses on the air space in between the two panes of glass. To significantly decrease the effect of convection in double-glazed windows, the double glazing is filled with gases that are less mobile, such as krypton, carbon dioxide, and argon. Another way to cut back on convection is to limit the space in which gas can move; without as many places to go, the gas has no choice but to keep still.

Another method to reduce convection is to create a vacuum,

in which convection currents are unable to alternatively ebb and flow. The problem with this approach is two-fold: first, the vacuum must be able to last the entire life of the window — probably around 50 years — and secondly, the vacuum required to neutralize convection's flow is far too large for most windows.

If the vacuum method is used, the window in question will need strong internal spacers for its structure, which will prevent it from collapsing inward due to the external air pressure. However, the structural spacers conduct back and forth between the gaps, resulting in a slight decrease in performance.

A far simpler solution to most of the aforementioned is to insert two or more extra panes of glass between the double-glazed panes. As an alternative, transparent sheets of plastic film can be inserted in lieu of extra window panes.

Because transparence is important for windows, many transparent forms of insulation are being researched. One is a transparent type of plastic that is filled with bubbles of insulating gas. The bubbles are trapped to reduce their movement. This and other forms of transparent insulation could one day revolutionize the world of windows, but at present, they are lacking protection from the negative effects of light and weather.

The final key to understanding the transmittance of heat from place to place is radiation. Just as the sun radiates its heat onto the earth, heat energy can be radiated elsewhere. The quantity of radiation fluctuates, and is determined by the radiating body. Building roofs radiate heat into the atmosphere, for example. Some surfaces have very low emission ratios, meaning that they will radiate only a tiny amount of their heat outward.

2.5: SOLAR COLLECTORS

To store and subsequently put the sun's energy into action, solar collectors are used to capture sunlight before converting it to heat. Solar collectors vary in shape and function. Flat plate water collectors typically have single layers — sometimes double — and are reinforced with plastic. Their extremely black surfaces result in high solar absorption. Heat pipes are hollow tubes that are filled with liquid at a specific pressure to instigate boiling at one end of the tube; vapor condenses at the opposite end. Heat pipes are capable of creating extremely large amounts of heat. Other types of solar collectors are available as well.

The difficulties and requirements of maintaining solar collectors are intertwined. Roof-mounted collectors are usually rather difficult to reach for maintenance, and they must be firmly attached to the roof to avoid leaks, which can throw off the collector's functioning. Because they exist outdoors, roof-mounted collectors must also be constructed to be impervious to Mother Nature's assortment of heavy winds, rainfall, frost, hail, and other potential weather conditions. Roof-mounted solar collectors must also be constructed and placed to avoid internal corrosion.

2.6: TYPES OF SOLAR ENERGY

There are several types of solar energy, due to there being so many different ways to apply it. Each of the following types has strengths and weaknesses, and you will find that more often than not, each overlaps with the other in some capacity.

Active solar systems use mechanical components, such as fans and pumps, to capture sunlight and process it into electricity or heat. In addition, active solar systems can be used to store heat for future use, as well as cause air movement for processes such as heating and, perhaps somewhat surprisingly, cooling.

As mentioned, the uniqueness of active solar systems is that they rely on mechanical components to function. This can be a great benefit, as active solar systems equipped with tried-and-true technology often result in a phenomenal solar savings fraction, which is the amount of solar energy provided by the solar technology divided by the total amount of required energy.

Unfortunately, the mechanical workings of active solar systems prove to be a double-edged sword, as their use results in operating costs — equipment must be properly maintained, after all — and the emission of greenhouse gasses.

Contrary to active solar, passive solar systems make use of non-mechanical techniques to capture and convert sunlight into a usable, beneficial form. If constructed and situated properly, structures with passive solar systems can provide heating, cooling, ventilation, and lighting for inhabitants.

Due to their ability to shirk mechanical equipment, properly constructing a passive solar-powered structure can be quite difficult. It is not a matter of simply constructing a building with many windows and expecting the sun to take the bulk of the work; the sun will work for you, but only if certain requirements are met. A building intended to be powered by passive solar means should face the sun, and the building's interior must be designed to circulate air naturally.

It should be noted that in certain scenarios, passive solar systems can make use of mechanical equipment. The catch is that the equipment must be automatic.

Passive solar will be more thoroughly examined in section 2.6, "Passive Solar."

Direct solar technologies make use of a single-step procedure to convert sunlight into usable energy. Indirect solar technologies make use of procedures that involve multiple steps to accomplish the same goal.

2.7: PASSIVE SOLAR

When the Romans vanished into the realm of history books, they took an important technological advancement with them: the ability to create large sheets of glass, which were most commonly used for windows. As mentioned in section 2.2, the Romans popularized the installation of large windows to heat their bathhouses. The art reemerged in France during the late 17th century, almost to the detriment of the 18th and 19th centuries, which saw overcrowded cities and buildings that had not been optimized to take advantage of the sun.

Properly heating a passive solar building requires common sense above all else. Popular rooms, such as living areas, should face south, with less common rooms, such as bedrooms facing north. Avoid over-shading by constructing buildings at a good distance from one another, if possible. There is no point in designing a passive solar-powered building if a nearby building's shade will negate the sun's benefits.

Free heat gains, which are contributions made to a building's heating by way of natural activities — cooking, washing and drying clothes and dishes, lighting, and using appliances and body heat — are always of assistance in keeping the interior of a building warmer than the outside. Sunlight streaming through windows is helpful, and though their prolific use should not be encouraged, using a traditional, fossil-fuel powered heating system also contributes to a building's heat.

Another excellent source of free heat gain is the window. For buildings that face the proper direction, a large window — the larger the better — will be a constant source of sunlight and heat.

It is important to use copious amounts of thick insulation to retain the structure's heat. Like a poorly sealed jar, heat can leak out of a building with insufficient insulation, rendering the collecting of heat almost useless. A solar collector is not much good if you will not be able to use the heat you collect later.

Not only is it important to design passive solar systems to use heat, but they must be designed to *not* use it as well. Buildings designed too optimally for heat will cause distinct unpleasantness during warmer months.

2.8: INTERSEASONAL STORAGE

During warm seasons, converting the sun's energy into heat will not be nearly as important as it will be during winter months. While winter does see its fair share of sunlight, certain measures can be taken to store spring and summer sunlight for use during autumn and winter. Unfortunately, this is an expensive tactic.

To save up the large amount of energy needed for a lengthy winter season, a hot-water storage device will need to be used. However, the hot-water storage unit will need to be at least as large as the building it is trying to heat. In addition, the hot-water storage unit's insulation will need to be half a meter thick at a minimum in order to retain the energy from summer until winter. The insulation is important because, just as with our discussion of passive solar systems in section 2.6, garnering a surplus of solar energy will not do any good if it can not be saved for future use.

2.9: DAYLIGHTING

Because the sun is only in the sky for approximately 12 hours a day in many locations, it is important that you get as much use from it as possible; that is the goal of daylighting. As discussed, certain measures, such as properly facing buildings and using large windows, further this goal. Along this line, a room's architecture should allow sunlight to flood it as thoroughly as possible; massive bay windows are not much good if pseudo-fancy outcroppings block solar energy's objective of providing heat and light.

If possible, try to construct a light well in the center of a building, in addition to installing lights on the roof. Also, the larger your windows, the more light will permeate the building, so use as much glass as you can. Skylights are capable of ushering in more light per area than windows, and they spread it more liberally over any given space. A word of warning: beware of stratification, which is the tendency for warm air to become trapped in a skylight well. Stratification leads to increased heat loss in cool climates.

As an alternative to skylights, some prospective builders may choose to use light tubes, which are cylindrical tubes made of highly reflective materials that are inserted into a structure's roof. Like a showerhead dispensing water, light is "sprayed" out of the bottom segment of a light tube after sunlight is reflected from its top.

Some experimentation with movable mirrors has been conducted, enabling sunlight to be "tracked" and more properly used should windows be poorly placed. Other experimentation with optical fibers and light ducts has taken place as well. Keep in mind that such devices can still be considered "passive solar" so long as their use is automatic, thereby rendering manual control unnecessary.

Eventually, the sun will set, and artificial lighting will need to be used. In such cases, energy can still be saved by only using artificial lighting when necessary, such as during naturally dark hours. Optimally, automatic artificial lighting systems, such as lamps that click on at a certain time, should be used to make sure all daylight hours are used effectively.

2.10: LOW TEMPERATURE APPLICATIONS

As mentioned, solar energy's most common application is providing heat. Several different methods for achieving this objective have been discussed, though primarily only within the realm of passive solar systems. The following subsections discuss applications for solar energy that require low- to mid-level heat.

2.10.1: Domestic Water Heating

Domestic water heating is believed to be the best application of active solar heating in all of Europe; heating water accounted for approximately 7 percent of the total national energy usage in the UK during the year 2000. Typically, incoming water is at a temperature close to that of soil, and it needs to be heated to around 60 degrees Celsius. The temperature can be slightly lower, but only slightly — the lower the temperature, the more prone the water becomes to harboring bacteria, such as Legionnaires' disease, which can result in pneumonia.

Domestic water heating is most commonly handled in one of a few ways. The first method is through common electricity, via an immersion heater in a hot-water storage cylinder.

Secondly, through an "instantaneous" heater, the power for which is usually provided by electricity or gas. Finally, just as in the first method, a hot-water storage cylinder is used, but it will contain a heat exchange coil, which is connected to either a district heating supply system or a central heating boiler, which is usually gas-fired.

Problems do exist with domestic water heating, all of which are traced back to heat loss. If the hot-water cylinders are not constructed with sufficient insulation, or if the pipe work is shoddy, heat can escape.

2.10.2: Space Heating

Providing heat to most structures is logical, as human beings and other forms of life require warmth to survive. Too much warmth, however, causes overheating, which is something we want to avoid.

The goal of space heating is to warm an interior to approximately 20 degrees Celsius. Certain times of the year, known as heating seasons, provide the best opportunity to accomplish this goal. Given the vastness of planet earth, it makes sense that different continents experience heating seasons at different times of each year.

It is important to consider not only location, but climate as well. Certain places will only be truly receptive to solar energy during certain times, such as during sunny winter months. As mentioned previously, it is vitally important to consider both location and climate when planning to make use of solar energy.

2.10.3: Active Solar Space Heating

If possible, it is probably best to use a passive solar system if your intent is to heat a structure, as active solar space-heating systems have quite a large failing: waste. Just like using a hot-water storage device to save a summer's heat for use during winter seasons (discussed in Section 2.7), an active solar heating system's collector will need to be quite massive, which will more often than not result in not all of the stored heat being used.

An attempt at negating this failing of active solar space heating is to make sure that the structure in question is extremely well insulated.

2.10.4: Other Ways to Minimize Space Heating's Demand

As mentioned, proper insulation is the best way to keep heat inside where it belongs. A way to help accomplish this objective is to "superinsulate" a house by using 200 or more millimeters of insulation.

Also, proper glazing on a window is instrumental in making sure as much sunlight as possible permeates the appropriate space.

2.10.5: Types of Heating Systems

Just as there are many different ways to meet low temperature heating demands, there are many systems available to meet demands.

A direct gain system uses windows to absorb the sun's rays and heat the interior. Solar energy gained via this manner is quite useful, unless a building happens to be improperly designed. In such cases, overheating is almost guaranteed to occur.

A conservatory, also referred to as a sunspace, is a greenhouse that is attached to the rear of a building. Heat is collected within the conservatory and subsequently transferred into the attached building. When new buildings are constructed to take advantage of solar energy, they are commonly built with direct gain benefits in mind, such as using large windows, light wells, and other newer technologies. Conservatories are most beneficial for older buildings that were not constructed to take advantage of solar power.

Adding on a conservatory is rather expensive and should not be justified on only the energy savings they will bring. While the performance of some conservatories can result in savings of 800 kilowatt hours (kWh) a year, not all of that savings is solar energy. Given that a conservatory acts as extra insulation to the building's south side, 15 percent of the 800 kWh is due to the building's thermal buffering. Another 30 percent is due to conduction through the intervening wall, which allows solar gains to enter the building through the conservatory. Finally, a whopping 55 percent comes from preheating all ventilation air making its way to the house. The fresh air entering the building via the conservatory will be warmer than air entering another way.

It is advised that you only build a conservatory if you understand that the conservatory will only save energy if it is not heated similarly to all other areas of the building. It is not merely just another room of the building, and it should not be connected to a central heating system, should one exist. Doing so will result in a guaranteed negation of the benefits of having a conservatory.

One way to cut the cost of adding a conservatory onto a building is to integrate the sunspace into the design. Just because older buildings are the primary adaptors of conservatories does not mean that new structures cannot take advantage of them. If possible, take advantage of earth sheltering, which sees the rear of a building built up with the earth, thereby bestowing extra thermal protection from the worst winter has to offer.

An alternative to conservatories is the Trombe wall, which is a wall facing the sun that has been constructed from thermal mass materials such as metal, stone, water tanks, concrete, or adobe. The sun-facing wall combines with insulated glazing, vents, and air space to form a large solar collector.

Radiation warms the wall and heat is sent into the house from the inner side. On bright, sunny days, the air circulates through the air space and into the house behind the wall, but at night, as well as on cold days, the air flow is completely cut off.

The Trombe wall is quite malleable in terms of its design.

- Windows can be placed within the wall for natural lighting, or even superficial aesthetics, but those who choose to implement this strategy should expect lowered efficiency rates. However, if the outer glazing has a high level of ultraviolet transmittance, and if the Trombe wall's window is made of normal glass, ultraviolet light can be used to provide heating while simultaneously protecting material objects and, most importantly, people, from ultraviolet radiation. This is in direct comparison to windows containing high levels of ultraviolet transmittance.

- If controlled by thermostats, electric blowers can be used to improve heat and air flow.

- Some Trombe walls make use of a trellis, a construction of wood typically used to support a climbing plant, to shade the solar collector during brutally warm summer months.

- At night, insulating covering can be placed overtop of the glazing surface.

- Shades, either movable or fixed, can be used cut down on nighttime losses of heat.

- If exhaust vents are placed near the top of the Trombe wall, and are opened to vent into the outdoors during the summer, the Trombe wall will pump fresh air through the building during the day, regardless of whether a breeze exists.

- Fish tanks can be used as thermal mass.

2.11: PHOTOVOLTAIC

Arguably one of the most effective uses of solar energy — and inarguably one of the most expensive — is the use of photovoltaics (PVs), which are used to convert sunlight directly into electricity.

2.11.1: History of Photovoltaic Technology

A French physicist by the name of Alexandre-Edmond Becquerel is credited with the finding of what would come to be known as the photovoltaic effect. In 1839, Becquerel published a paper in which he described his various experiments with a wet cell battery. The paper documents his eventual discovery that the battery's voltage underwent a significant increase when its silver plates were directly exposed to the sun's light.

Almost 50 years later, an electrician based in New York named Charles Edgar Fritts put together a solar cell made from selenium. The modern solar cell is very similar in design to the one made by Fritts in 1883. Fritts' selenium solar cell was made of a thin wafer of selenium coated with a grid of slender gold wires and a sheet of glass. Unfortunately, Fritts' design produced an efficiency rating of less than 1 percent, but it was a step in the right direction.

PV technology developed at a steady rate until an explosion of progress occurred in the 1950s at Bell Telephone Laboratories in New Jersey. Calvin Fuller, Gerald Pearson, and Darryl Chapin, deservedly labeled as pioneers in the realm of PV development, produced pieces of silicon that were far more efficient at deriving electricity from light than earlier advancements. Fuller, Pearson, and Chapin used a process called doping, where the properties of a semiconductor are intentionally changed due to an introduction of impurities into the semiconductor.

In 1958, the successful application of solar cells powered a radio transmitter in Vanguard I, the second U.S. space satellite. This successful venture has resulted in PV cells becoming omnipresent as a spacecraft power source.

2.11.2: Potential

Theoretically, the use of PV technology could easily provide more than enough electricity to power the entire U.S. If we were to construct PV panels that would cover 14,000 square miles — which equates to slightly less than 15 percent of the state of Nevada — our entire country would have enough electricity to last for quite a while.

Of course, I say "theoretically" for a couple of good reasons. First, devising such a system is one thing, but implementing it is quite another. Because PV technology is reliant upon sunlight, even the largest grid is doomed to click off at night. Secondly, the sometimes exorbitant cost of PVs prohibits them from being as ubiquitous as many would like.

Simply put, anywhere the sun shines is prime real estate for PV technology. Even on the cloudiest day, the sun provides more than enough light to be converted into electricity. In fact, out of any of the renewable forms of energy in existence — most of which you will learn about in this text — PV technology is the least restrictive.

Unlike many renewable energy sources, PVs have no moving parts, which results in a significant decrease in noise pollution. Different PV technologies have resulted in PVs that can be installed for use anywhere. PVs can be used to power a single light, a home, a block of homes, or an entire community.

2.11.3: Types of PV Technology

All PV technologies fall into two modules. The first is crystalline silicon, which you can find in most PV commercial-use systems. These produce electricity with the intent of selling it, rather than using it directly.

Silicon starts as sand, which is then mixed with a minute amount of a substance containing various electrons, such as phosphorus or boron. When light makes contact with the PV material, the electrons from the first step are dislodged. The dislodgement of the electrons results in an electric current.

The conversion efficiencies of crystalline silicon modules are reasonably high, falling within the 12 to 14 percent range. Another boon is that crystalline silicon modules are made from pre-existing materials, most commonly from chip-manufacturing plants that discard silicon. Unfortunately, crystalline silicon modules can be extremely difficult to manufacture, and their price is quite high.

The second type of module is thin-film, which has the advantage of being able to produce electricity from very thin films, but otherwise maintains the same foundations as crystalline silicon. Because of its advantage, thin-film is often integrated into materials such as building tiles. Due to the thin-film modules being nothing more than a part of the tile, the modules are barely able to be seen, and thus are not a bother to those who regard constructions such as PV plants as visual pollution.

Besides materials such as building tiles, thin-film has seen a near-seamless integration into other, more commonly noticeable materials. Amorphous silicon can be found in watches and solar calculators.

Thin-film PV modules are generally easier to manufacture than crystalline silicon due to requiring much less material. Because of this, thin-film PV modules are commonly produced large-scale, rather than only a few at a time. As a trade-off, thin-film PV modules have somewhat lower efficiencies. However, this is expected to change as thin-film technologies undergo almost constant advancements.

2.11.4: How do PV Modules Work?

PV modules, which are sometimes referred to as panels, can work individually or be amalgamated into arrays. A tiny array consisting of only two panels can be used to power a residential electricity system, or thousands upon thousands of modules can successfully, and easily, power a system of 100 kilowatts (kW) or more.

Modules make up the core of any PV system, but other components are fundamentally necessary as well. Balance-of-system components, such as a step-up transformer to increase the voltage, an inverter used to convert the direct current produced by the module to the grid's alternating current, and simple but oft-forgotten materials, such as supports, miscellaneous hardware, and brackets, help maintain PV modules and are typically responsible for about one-third of a system's total cost.

2.11.5: Cost

In direct opposition with their phenomenal abilities, PV

technologies also have extremely high costs. Finding the cost of a PV arrangement is done by calculating the dollars per watt. This will determine only initial costs; factors such as maintenance, operating costs, and local sunlight levels will help determine a system's true costs.

As improvements to PV technologies are made, costs will continually decrease. Large-scale systems generating more than 100 kW are dramatically cheaper than smaller ones, such as those used to provide electricity to residences.

Location is another important factor in determining cost. Areas with fewer periods of strong, direct sunlight will see higher per-kilowatt-hour costs, whereas areas with almost constant sunlight will see lower per-kilowatt-hour costs. Also, because of technological advances, modern systems typically produce lower costs than those produced years prior.

2.11.6: Photovoltaic Power Plants

When electricity is needed on a substantial scale, the common tactic is to construct PV power plants by way of centralized, ample PV power systems. Such power plants have been known to provide electricity in Switzerland, Germany, the U.S., and Italy.

The ability to purchase and install large numbers of PV modules — all of which can be located on a site partial to heavy solar radiation — is one major advantage of large, stand-alone PV plants. One major drawback may be that the electricity produced on-site is not used on-site; it must be distributed via the grid, which often results in heat losses, and subsequently allows for the electricity to be sold only at wholesale prices. In some countries, such as Germany, electricity is usually purchased at a premium.

Land preservationists are unhappy about the size of most PV power plants, but fortunately, the land can often be used for purposes other than just the power plant. Kobern-Gondorf in Germany is rather large, but its land is also used as a nature reserve to provide shelter for a variety of flora and fauna.

PV plants located in prime solar radiation areas are far more successful than those located in areas such as northern Europe are. Areas in southern California reach solar radiation totals more than twice those that areas in Britain reach, and they have clearer skies. In areas such as southern California, annual energy output is also increased due to solar radiation being more direct.

2.11.7: Strengths and Weaknesses

As already discussed, PV technology comes pre-packaged with benefits, such as noiselessness, and diminishes visual pollution due to its ability to be placed on rooftops or integrated into pre-existing materials, such as roof tiles, calculators, watches, and more. Also, because PV modules can be sized and fitted to almost anything, they are able to be directly applied to devices and structures that have need of electricity.

Unfortunately, their reliance on the sun is not only their greatest strength, but also their greatest weakness. On an overcast day, PV-generated electricity cannot be relied upon. It will still be produced, but not in the capacity it would on a clear, sunny day. Also, regions dependent upon electricity at night will probably want to make use of an automatic lighting system to make up for PV's lack of nighttime output.

2.11.8: Environmental Issues

PV technology is unique in that it does not feature environmental drawbacks nearly as harmful as other forms of renewable energy, though some do exist. Fortunately, most can be prevented with careful construction and common sense.

Some modules contain very small amounts of toxic substances. Should a fire break out within a PV array, traces of the substances will be released into the environment via the fire's smoke. Also, as with any type of electrical equipment, there is the possibility of a person receiving an electric shock. This possibility becomes more likely as systems grow in size, but as long as the PV system is well-engineered and properly maintained, any shock should not be any more harmful than that from comparable electrical stations. Even so, it is important to remember that the possibility of a shock does exist, and no matter how small it might be, getting shocked is never an enjoyable experience.

Other toxic substances, such as traces of cadmium, can be found during the manufacturing process in PV modules. In this and any instance of toxic chemicals that have the potential to be released, developers must make sure that PV plants are carefully designed and operated. The more precautions that are taken, the less likely it is that anything dangerous will be emitted should a plant malfunction occur.

Finally, although PVs have incredibly long life spans, they will eventually become garbage. Instead of throwing them out, however, make an effort to recycle them.

POWERING MR. FUSION - BIOMASS AS AN ENERGY SOURCE

3.1: INTRODUCTION

During my youth, one of my favorite movies was Back to the Future II. As a young boy, I was interested in the film's vision of what I would have to look forward to in the year 2015: holographic movie trailers, self-drying jackets, hover boards, pizzas ready for consumption in less than one minute, and, of course, flying cars.

To me, the De Lorean DMC-12 vehicle was the most interesting aspect of the film. I recall my fascination as Emmett "Doc" Brown flipped down his shades, gripped the steering wheel, and said "Where we're going, we don't need roads." Seconds later, the De Lorean took to the air and disappeared in a brilliant display of special effects.

Though a fantastic scene, perhaps the most intriguing facet of the De Lorean's futuristic makeup was its fuel system, "Mr. Fusion." The Doc rummaged through scraps of garbage in order to provide power to the car, and as I watched, fascinated, I thought of all the bickering I had ever heard my parents make about constantly rising gasoline prices, and how with this system, they were irrelevant. Who does not have garbage to dispose of, after all?

Though I did not realize it during my childhood, biofuels such as solids, liquids, or gases that consist of, or are derived from, biomass were not as far off as the year 2015.

3.2: A HISTORY OF BIOMASS

Hardly a science-fiction source of renewable energy, biomass has existed since humans first learned that burning wood could keep them warm — and even before that. Our ancestors burned wood to keep themselves warm for centuries, used animal fat to construct tallow candles, and used grass to "fuel" their primary method of transportation: horses.

The dependency on biomass as a fuel source existed until approximately the Industrial Revolution, at which point fossil fuels, such as coal, natural gas, and oil, became many countries' primary means of heating homes, powering transportation, and much more.

Despite relatively low usage, compared to more modern technology in countries such as the United States, approximately one third of developing countries still maintain a heavy reliance on biomass as their principal means of energy consumption. Though the usage is ever decreasing as countries continue to develop, it is possible that history might be forced to reverse itself. Fossil fuels are depleting quite rapidly, which could eventually necessitate a reliance on biomass and other renewable energy sources, instead of the other way around.

3.3: WHAT IS BIOMASS?

Whether driving around in a car, walking on foot, or flying through the air, the chance is good that you will see something eligible for conversion into biofuel. Biomass may consist of almost any biological materials, including wood scraps and sawdust left over from lumber processing, forest and crops grown specifically for biomass, non-edible portions of crops left behind after a harvest, tree stumps, plant remains, forest residue, sewage and farm-animal waste, and agricultural scrapes such as pits, plant stalks, shells, and other pruning remnants. The possibilities for materials that can be successfully converted into biofuel are nearly limitless.

3.4: THE CONVERSION PROCESS AND ITS PROS AND CONS

The technological knowledge for morphing biological materials into useful biofuel has existed for quite some time. Consider the steam boiler, a useful mechanical marvel that has been in operation for years. It boasts incredible reliability, inexpensive use, and low maintenance costs.

Using a process known as direct combustion, biomass materials such as those listed above are burned directly in a steam boiler, undergoing the exact same process as coal. The heat resulting from the burning fuel converts water into steam, which in turns drives electricity generators.

This form of electricity production is a great alternative for a number of reasons, both practical and legal. First, waste takes up a large amount of room. Getting rid of this waste via adaptation into biofuel not only gets rid of intrusive garbage, but converts it into something beneficial and necessary to the functions of the human race. Second, landfills are quite expensive to maintain, making direct combustion a lucrative alternative to maintaining piles of otherwise worthless trash. Finally, burning is typically considered illegal unless permission is obtained to do so; burning for biomass, however, is officially permissible for large-scale facilities in possession of the proper documentation.

For all the advantages and benefits of direct combustion, it is not without problems. Fuel variability dictates that some fuels will be moister than others, and they will contain fluctuating amounts of contaminants, in addition to other concerns. Though most direct combustion systems are tolerant of a fuel collection's diverse assortment of variances, any given collection's numerable variations will often hinder plant efficiency, as well as the ability to control emissions.

A study conducted in a 1998 edition of Financial Times Report revealed that most biomass fuel systems, especially those using direct-firing technology — which is used in direct combustion systems — operate at as low as 14 percent or as high as 18 percent efficiency rates when pitted against comparable fossil fuel systems.

3.5: METHODS FOR IMPROVING FUEL EFFICIENCY

Despite low efficiency rates, all hope is not lost for fuel efficiency, and a number of tactics exist to provide improvement.

In terms of direct combustion, time-consuming yet somewhat simple solutions are available. Fuels can be sorted into different categories, primarily based on moisture, before burning. This tactic eliminates many systems' difficulties in dealing with fuel variability, one of the main problems with raising efficiency levels. Also, many fuel sources can be dried before the conversion process begins, thereby reducing some moisture and perhaps even eliminating the need for sorting. Finally, equipping boilers with larger steam turbines will extract more heat, given that larger turbines possess additional cycles.

An alternative process known as gasification exists, which sees fuel converted to a combustible gas and subsequently used to drive a turbine directly. Gasification is a technology that comes with extremely high efficiencies, but is not widely used. Many commercial systems simply are not available, though some processes, such as landfilling, do make strong cases for gasification becoming mainstream.

A final suggestion sees the complete elimination of a pure biomass turbine. The procedure of co-firing dictates that a small amount of biomass fuel, typically between 5 and 10 percent, be added in with coal in the world's many coal burning plants. In addition to eliminating biomass turbines, co-firing also significantly reduces the emission of nitrous oxides, such as sulfur. Co-firing a mixture of biofuels and natural gas increases the fuel's heat content, thereby improving both equipment performance and combustion.

It is difficult to say that any one of these methods is the best. Sorting and drying fuels takes time and labor, but it does see an increase in efficiency levels. If landfilling and other gasification processes were more commonplace, efficiency levels would soar dramatically, but this process, like all the others, comes with setbacks.

It is important to carefully research the available solutions, as in the end there is no "best" answer available. The solution will largely depend on a given individual's or company's goals and capabilities.

3.6: CROPS

Rising from bed to greet the morning sun, feed the animals, complete household chores, and tend to crops will be familiar to anyone who grew up on a farm, but in terms of biomass, an integral undertaking is conspicuously absent from that list: gathering crop wastes for biofuel conversion.

After crops are tilled, hoed, picked, and gathered, many valuable resources are left behind. Consider wheat and corn, the residues from which can be used as bedding or even feed. In major cereal-growing regions, more than half of these and other helpful crop residues go unused. Why not apply the wastes to something as useful as energy renewal?

Just as useful — and needlessly wasted — is fibrous sugar cane residue. Commonly used in sugar factories as a steam-generating fuel, the residue from sugar cane is largely responsible for providing the electricity that powers such plants. The costs required to transport sugar cane residue to sugar factories is rather prohibitive, but the trade-off is the generation of electricity, sometimes in excess, and is something to consider. An electricity surplus can outweigh transportation costs in some, but not all, cases.

3.7: OILY SEEDS AND BIODIESEL

Though not wasted as often as wheat, maize, and sugar cane residues, oily seeds are another seemingly one-dimensional product that holds potential as a renewable energy source. Sunflowers, oilseed rape, and more are grown almost specifically for the oil contained in their seeds, which is able to be converted into biodiesel, a diesel-equivalent fuel that can be used individually or in conjunction with conventional diesel fuel in unmodified diesel engines.

In addition to being completely biodegradable, biodiesel is nontoxic and commonly produces 60 percent fewer emissions than traditional, petroleum-based diesel fuel. Unfortunately, biodiesel comes packaged with approximately 35 percent greater hydrocarbon emissions, which are responsible for forming smog.

Despite reservations, many vehicle manufacturers are optimistic about biodiesel, claiming that it is better than standard diesel due to its ability to clean engines by removing fuel line deposits. Outside the obvious realm of automobiles, biodiesel can also be used as a heating fuel in many commercial and domestic boilers, though potential adapters are cautioned to refrain from burning B100, pure biodiesel, in domestic heaters until first slowly increasing the amount of biodiesel used in these appliances, as B100 can sometimes dissolve heating oil, which can break off in chunks and cause malfunctions.

To obtain the oil for biodiesel from oily seeds, the seeds have to be crushed. The energy content found in oily seeds' precious resource is only slightly less efficient than traditional diesel, and, as mentioned, it can be blended with traditional diesel fuel. Unfortunately, it is not common for just anyone to have access to oily seeds, and large-scale collection and usage logistics will probably restrict the use of biodiesel primarily to large-scale functions, such as catering operations, rather than prolific home use.

3.8: BIOGAS

In the case of biomass as a renewable energy source, one man's garbage is literally another man's treasure. For years, human beings have been polluting the earth by dumping wastes in enormous piles around the earth. Though landfills are arguably not one of our better ideas, these mounds of trash can be put to a productive use — but not without prerequisite consequences.

Biogas is a form of biofuel produced by the breakdown of organic material, such as biodegradable waste, manure, sewage sludge, and municipal wastes. The breakdown process is commonly anaerobic, meaning it occurs with an absence of oxygen.

Perhaps the most common form of biogas is landfill waste. According to the Environmental Protection Agency, the United States contains more than 500 landfills that would suffice as prime sites for electricity generation via the methane gas generated from landfills. The procedure to attain and harness the nastiness from any of these pliable locations is relatively simple. First, the entire area must be covered with a material that is unyielding and largely impervious to harm. From there, pipes already rooted into the ground apply a vacuum to suck up the gas, which is either trapped and dispersed back into the landfill, or pumped to a sewage plant for appropriate treatment.

This process of gasification yields several positive direct use applications, and the recovered gas can typically be applied to anything for which natural gas is used. As previously mentioned, landfill gas can be used to generate electricity, which is arguably its most common use. Not only can the electricity be immediately used, but a surplus, if and when available, could easily be sold in order to power offsite homes and businesses. In addition, landfill gas can be properly converted to power furnaces and boilers, as well as kilns. Although landfill gas does have low energy conversion efficiency, landfills typically take a great deal of time to deplete their large stores of gas, which can lead to the gasification process being financially profitable.

There are many serious risks involved in landfill gasification that deter many from its use. Primarily, the high concentration of methane gas often results in explosions; the bigger the landfill, the greater risk for this occurrence. Second, but of equal importance, the quantities of methane gas contribute greatly to global warming, an already serious problem. Third, anyone who has driven past a garbage dump will wholeheartedly agree that the stench is beyond reprehensible. Not only are our noses polluted by the stink, but the air is harmed as well.

Gober gas, a biogas generated from cow manure, is smokeless and odorless after being properly treated and converted, and therefore ideal for uses such as cooking.

Gober gas is most commonly found in India, where it is generated at numerous micro plants attached to homes. A gober gas pit consists of a circular pit made of concrete, and it must be airtight and come with a hose connection. Manure is fed into the pit and added to a specific quantity of water. The gas pipe runs from the pit and into the home's kitchen fireplace via control valves.

Though not as prevalent as landfill and gober gas, other forms of biogas include swamp, digester (a plant, typically anaerobic, that treats certain types of waste), and marsh gas. A biogas-powered train, detailed in section 3.11, has been operating in Sweden since 2005.

3.9: COSTS

With all knowledge comes power, and with biomass, as

with anything worthwhile, comes a price. The amount of money required to operate a biomass facility successfully is largely dependent upon the facility's scale. Production facilities located near or on industrial sites, such as a lumber mill, have relatively low costs. These facilities have access to fuel that is readily available — after processing the lumber, simply round up all the residue and scraps — and it is essentially free. The lumber company is working with the wood, and now they can put its remains to good use. Doing so not only results in free biomass fuel, but it can often surrender negative net costs for the industrial facility if burning helps reduce disposal costs.

Last but not least, the efficiency of the fuel is already known. Again using our lumber mill as an example, all of the wood is gathered, sorted, and likely dry, and it is all readily available in one accessible location.

At the opposite end of the biofuel spectrum, facilities that do not have easy access to fuel face substantial costs. Given that the fuel is not located on-site, preparation, labor, and machine power must be sacrificed in order to collect, transport, and process the fuel into a useful form. One of the largest boons in the above lumber mill example was that the fuel's consistency was known due to its location, and this will probably not be the case with offsite facilities. Fuel size can range in different degrees of moisture, size, and other factors, which means greatly increased costs for the labor and energy needed to locate and collect the prospective biofuel.

Home consumers and businesses often feel the causality of high costs acutely, as biomass power plants will typically charge higher rates for the electricity they generate due to the expenditures they made in order to acquire it.

3.10: ETHANOL

Although chemists will rightfully claim that it is a combination of different compounds, ethanol is, at its core, a substance most people likely know better as alcohol. To create alcohol, we must make use of a process known as fermentation.

Fermentation is an anaerobic biological process that entails the conversion of sugars to ethanol. The breakdown process is driven by microorganisms, with yeast being the most commonly used in the process.

Because sugar is instrumental in the process of fermentation, it might seem obvious that the sugar most often used is sugar cane, though other options do exist. Plants such as corn, potatoes, and wheat contain starch, which can be converted to sugar. However, the process for converting starch to sugar, while not arduous, is more time consuming, given that starch is not the intended end result. Because the U.S. boasts corn as a plentiful crop, it is common practice to convert starches to sugar before dabbling in fermentation.

Wood can also be converted, but a good amount of pre-treatment is involved, and undergoing the aforementioned pre-treatment processes always requires a great deal of expense.

After fermentation is complete, a liquid containing 10

percent ethanol is our reward. The ethanol is then distilled, which is a purification process that involves the liquid being vaporized and subsequently condensed. The distillation process requires tremendous amounts of heat, which are typically supplied by discarded crop wastes. Think of it as biomass's own circle of life: a biomass fuel is burned to create another source of biomass fuel.

Although ethanol is not suitable as a substitute for petrol, it is eligible to be used as a gasoline extender in gasohol, which is petrol that contains 26 percent ethanol. Percentage amounts are rising as more advanced techniques for ethanol are discovered and subsequently put into practice.

In Brazil, there exists a program known as Programa *Nacional do Álcool*, or the National Alcohol Program. This program produces ethanol from the residues of sugar, and it is the single largest commercial biomass system in the world. The establishment of the program came at a crucial time in 1975, when oil prices continued to soar and the price of sugar was extremely low.

During its first 25 years in existence, the program produced savings of approximately $40 billion; further savings came from a reduced rate of interest on foreign debt. Estimates made in the year 1999 point out that vehicles running on gasohol's mixture of petrol and 26 percent ethanol was very helpful in reducing annual greenhouse emissions, all thanks to this tremendous program.

Economically, the production of ethanol tends to fluctuate, sometimes wildly. If the price of oil is up at the same time that the price of sugar happens to be down, then production is considered viable. Disagreements in Brazil between distillers, growers, and the national oil company have had an effect on ethanol production, which saw costs dip and rise constantly during the 1990s.

As mentioned, ethanol use in the U.S. tended to be high due to our abundance of corn crops, but at many points, cheap gasoline neutered ethanol production benefits. As the twentieth century ended, continued threats of high oil prices and a huge surplus of corn led to an ethanol boom, and the U.S.' liquid biofuel output has grown at a steady 5 percent per year.

3.11: THE BIOMASS-POWERED ENGINE THAT COULD

A good, constructive example depicting biomass in motion is the completely biofuel-powered trained, owned by Svensk Biogas AB.

Though in operation in Sweden, the schematics and successive construction of the train, running under number 1334, took place in Italy, and it was built by Fiat (Fabbrica Italiana Automobili Torino; in English, Italian Automobile Company of Turin) S.p.A., an Italian company specializing in engines and automobile design, as well as finances and industrial design. Fiat is located in Turin, Italy.

Train number 1334 is one of 100 trains sent to operator SJ, and 1334 is one of only 12 trains to feature specialized cargo compartments in each car. Train 1334 runs all the way from the north to the south end of Sweden on non-electric rails. In 1994, the train underwent a major overhaul that entailed brand new gearboxes and engines, as well as the rebuilding of the rail cargo compartment.

In addition to "under the hood" components, train 1334 also received some aesthetic and modern technology-based overhauls. Amplifiers for mobile devices have been installed, cutting down on the static that can sometimes accompany a train ride through rural terrain, and several wireless access points have been added for passengers who want or need to do some web browsing during their train ride. A large, flat-screen monitor has been installed in the railcar, perfect for watching movies, and hot drinks can be purchased from a vending machine. Finally, to make the train ride even more enjoyable, the ventilation system has been revamped to allow more efficient air circulation.

Pleasant aesthetics and modern superficialities are fine, but the most important facet of train 1334 is its reliance on biomass as a fuel source. Acting as replacements to standard diesel engines are gas engines with catalysts. Due to its new system, train 1334 emits absolutely no carbon dioxide into the atmosphere. Even better, the biomass fuel is produced locally. As mentioned, locally-produced biofuel results in significant cost reductions due to not having to travel offsite to collect, sort, and dry the source.

3.12: ENVIRONMENTAL ISSUES

Costs are not the most substantial issues biomass must contend with. Renewable energy sources come prepackaged with their own environmental concerns, despite usually being less harmful than fossil fuels.

Again falling back to our lumber mill example, wood burning can result in significant particle emissions. Sulfur emissions, for example, are generally less than what can be found in coal, but they are still very harmful, depending on the type of fuel being burned. Also, in relation to burning, some fuels not only smell horrible, but they may also contain unhealthy contaminants that are emitted during burning and sometimes during transport. In addition, animal wastes not properly converted and treated have the potential to pollute watercourses, making life difficult for both humans and local wildlife.

Speaking of wildlife, removing forests and other vegetation explicitly for biomass fuel is detrimental to the continued existence of any area's flora and fauna. Along a similar vein, many crop wastes are intentionally left behind and not converted into biofuels because of their ability to help with soil quality, keeping soil fresh and clean so that more crops can be planted and used for any number of purposes.

The failings of fuel variability and cost also need to be kept in mind. The smorgasbord of matter found in fuels that need to be sorted, dried, or undergo other operations makes designing and operating a high-efficiency biomass plant difficult, if not impossible. Fuels of varying moisture and contaminant content must be able to burn efficiently, but they may not be able to without being properly sorted. These and other operations require time, money, labor, and other resources.

GETTING WET AND WILD WITH HYDROPOWER

4.1: INTRODUCTION

Despite the sun's well-documented history of being regarded as a deity, water is perhaps the most vital ingredient to human life. Water is everywhere. We drink it, our bodies are largely composed of it, we wash dishes and clothes with it, we bathe in it, and we play in it; it is tasteless, yet it is always considered the perfect drink for any occasion.

Imagine standing on a beach and gazing out at the ocean during sunset. The horizon seems so far away, almost as though the water could stretch infinitely. It is impressive and humbling, especially from the perspective of your humble author who, as a little boy, unwisely disregarded his mother's instructions about keeping close to the shoreline lest the current snag him and drag him out to sea. Sure enough it did not take long for a wave to sweep my feet out from under me and drag me, coughing and sputtering, deeper and deeper.

My mother heard my cries and came sprinting toward me, dove into the water, and pulled me to safety. That was the first time I realized just how important it was to have respect for something so beautiful, yet so dangerous. Water is many things, but powerful should be foremost among any list. Water powers our bodies, but it also has the capacity to power a number of other things vital to our existence.

4.2: A HISTORY OF HYDROPOWER

Water is considered the earliest renewable energy source responsible for significantly reducing the workload of both man and beast.

Long before water distillers and bottled water manufacturers was noria, the device historians credit as being the first to be powered by water. The noria was circular and held a jar, probably made of clay, on each of its various platforms that jutted from the circle's exterior, similar to a sun's rays in a child's drawing. As a current aided the wheel in turning, each jar was dipped into the water and filled before being used for irrigational purposes. The noria's longevity is substantial: it evolved over a period of six centuries before the birth of Jesus Christ.

Before long, water wheel devices such as the noria led to civilization's first water mills. The earliest documented water wills were most likely vertical-axis mills, more commonly known as "Norse" or "Greek" mills. Vertical-axis mills first appeared during first or second century B.C., and originated in the Middle East. Some centuries later, Norse and Greek water mills also appeared in Scandinavia.

Not to be relegated to just methods of irrigation, the Greek and Norse water mills served several useful purposes. Their construction allowed them to grind corn, raise water, and keep excellent pace with several industries, such as coal mining, paper-making, iron working, wool, and cotton.

Contrary to Norse and Greek methodology, the Saxons were believed to have used vertical- and horizontal-axis mills, though the latter design was not widely used until the concept had ample time to evolve.

As humanity moved into the nineteenth century, even the most devout proponents of water-powered technology had reason to lose a faith. Steam engines fired by coal were becoming increasingly prominent. Fortunately, a young Frenchman named Benoît Fourneyron decided the time was ripe for water to regain its former glory. In 1832, Fourneyron came up with the first successful turbine, which was outfitted with guide vanes used to guide water's flow. Finally, after roughly 2000 years, the waterwheel had received a significant upgrade.

In addition to the implementation of guide vanes, Benoît Fourneyron's turbine brought with it a number of other innovations. The turbine could run while completely submerged, something unprecedented before that time. Fourneyron's turbine allowed water to be converted to mechanical output, and was able to achieve a conversion percentage of 80 percent. Instrumental in this phenomenal percentage was the turbine's rotor, which spun extremely quickly.

A short time later, James Francis, an American engineer, began conducting numerous experiments on inward-flow radial turbines. This design was contrary to the one Benoit

Fourneyron had created. With Fourneyron's turbine, water entered at the turbine's center before being direct across the curved faces of the guide vanes, enabling the flow to travel outward at a parallel to the runner blades' curve.

Just as before, Fourneyron's astounding invention saw little progress in hydropower, and stagnation occured shortly after James Francis' contribution to the cause. In September 1881, the quaint little town of Godalming in Surrey, United Kingdom, became the first town in the U.K. to be outfitted with a public supply of electricity. The source of power for this modern marvel was water. A traditional water wheel was used, and while the theory of the design was tried and true, the plant encountered one of hydropower's more common problems: the River Wey's flow was unreliable. Soon after, the water wheel was powered by a steam engine.

Research and subsequent improvement into hydropower surged for the final two decades of the 19th century, resulting in power plants that were capable of producing over a megawatt in capacity.

Currently, hydropower is responsible for supplying more electricity than any other single source of renewable energy. Approximately 10 percent of the U.S. is powered by H2O, and countries such as Norway, Sweden, and Switzerland receive the vast majority of their power from hydro sources.

4.3: HEAD

Of course, water coming from a higher elevation is "more

powerful" than water that does not come from quite as high. Water's elevation is known as its head, and it is typically divided into three different categories: low, medium, and high head.

A high head typically constitutes an elevation greater than 100 meters. The required flow for such a high head will obviously be significantly smaller, as the high head has gravity on its side. A low-head plant is generally considered to be less than or equal to ten meters, and of course, will require a greater flow.

Head is important, but there are a variety of other factors — such as volume and current, as mentioned in our example — to take under consideration. Consider the following scenario. Two different hydropower plants are in operation. Hydro A has an extremely high head (from the top of a mountain, for example) and a fast water current, but a low volume. Hydro B sports a very large volume of water, but a slow current and a medium- to low-sized head. These two plants would function quite differently.

4.4: TURBINES

As powerful as it is, all the water in the world will not be useful unless it has a powerful turbine to direct its traffic. Just as there are different types of head, so are there wide varieties of turbines that can be applied to any water-power system arrangement.

4.4.1: Turbines: A General Overview

No matter the name it is called, or the elevation it will be working with, every turbine consists of a set of curved blades. These curved blades are integral in deflecting the water in a way that will force the water to relinquish as much of its energy as possible.

Also crucial to a water power system is the runner, the rotation system responsible for spinning the blades around. The water makes its way to the runner by being sent through channels, a powerful jet, or a set of guide vanes, depending on the type of turbine.

Some water power systems can feature more than one type of runner. The Galloway plants located on the River Dee in Scotland's south-west region feature two runners: a Francis turbine — discussed in Section 4.7, "The Francis Turbine" — and a propeller-based runner. Propeller runners, also known as "axial flow" turbines, are most commonly found in systems that feature low heads with large flows. Francis turbines do their best work when used with medium or high heads.

In the case of axial-flow turbines, water enters in an area that is as large as possible; the more room, the better for this type of turbine. Because of this, axial-flow turbines are good for handling extremely large volumes of water. Compared to certain other types of turbines, axial-flow turbines have a distinct advantage in that should efficiency need to be improved, all that is required is some tweaking to the angle of the runner blades when a change in power demand takes place. "Kaplan" turbines are axial-flow turbines with this feature.

4.4.2: Types of Turbines

≋ **Kaplan Turbine:** With Kaplan turbines, the water is swirling around just as in an unaltered axial-flow design; the distinction is the speed of the blades, which is now significantly greater than the speed of the water, almost twice as fast. Because of this significant upgrade, a Kaplan turbine permits a high rate of rotation, even if low water speeds come into play.

≋ **Pelton wheel:** The Pelton wheel was developed by Lester Allan Pelton, an inventor by night, and a millwright and carpenter by day. Pelton's wheel, developed in the 1870s, is the byproduct of an accidental observation. While observing a water turbine, the piece responsible for securely holding the wheel onto the shaft slipped, resulting in the water flow changing to a half circle. Pelton observed that rather than detracting from the design's efficiency, the slip-up resulted in the turbine's speed going even faster, thus providing a dramatic increase in efficiency.

Though used throughout the 1880s, it was in 1887 that a miner fixed Pelton's wheel to a dynamo, the first electrical generator, which brought about the birth of one of the first hydroelectric power plants, used in the Sierra Nevada Mountains.

For heads equal to or above 250 meters, the Pelton wheel is considered the optimal, preferred turbine to use. Dominant in its execution, yet simplistic in its design, the Pelton wheel is a wheel mounted with a set of double cups around its rim. When a powerful blast of water forms under the high head's pressure, it hits the edge between every pair of cups as the wheel continues to turn. The water then makes its way past the curved bowls and, if conditions are favorable, releases all of its built up kinetic energy.

Should the Pelton wheel's power need to be tweaked, the size of the jet of water formed beneath the high head's pressure can be altered, thereby changing the flow rate's volume. Or, the stream of water can simply be deflected away from the Pelton wheel.

For those who desire maximum efficiency, the Pelton wheel's cup speed should be approximately half that of the water jet's.

≋ **Turgo turbine:** Bursting onto the scene in the early 1900s, the Turgo turbine has single, shallower cups, as opposed to Pelton's sets of doubles. The cups on Turgo turbines direct water in via one side, and right back out via the other. The water makes its entrance as a fast-moving, powerful jet stream, which strikes all the cups in turn; this labels the Turgo turbine as an "impulse" turbine.

Contrary to the design of the Pelton wheel, a Turgo turbine is able to handle a much greater volume of water easily, even when pitted against a Pelton wheel with an identical diameter. This advantage makes the Turbo turbine the perfect candidate for a medium head's power generation requirements.

≋ **Cross-flow turbine:** The cross-flow turbine is also known as both the Banki-Mitchell and Ossberger turbine. This turbine was developed by a trio of inventors including German Fritz Ossberger, Australian Anthony Mitchell, and Hungarian Donát Bánki Ossberger, who filed the cross-flow turbine's first patent in the early 1920s; in the present, modern world, Ossberger's company is considered the leading manufacturer of the cross-flow turbine.

The cross-flow turbine is also considered an impulse turbine. However, this particular design sees the water entering as a flat sheet, not as a jet. The water flows smoothly across the blades and around the central shaft before sliding across blades set on the opposite side before finally exiting the turbine.

The cross-flow turbine is widely used in plants of a smaller scale, contrary to the Francis turbine.

4.5: HOW DOES A HYDROPOWER SYSTEM WORK?

Just as there are different uses for water and three different categories of head, there are different methods for classifying hydropower systems.

- By the water head's efficiency.

- By the type of installation — whether a dam or a reservoir — in addition to its location.

- By a head's effectiveness. Is it a low head? Perhaps a medium? It is crucial to take the category of head into account.

- Finally, by the type of turbine that the system will use.

A standard hydropower system will include a dam, a healthy water reservoir upstream of the required dam, a turbine, a generator, and a penstock, which is responsible for carrying water from the reservoir to the turbine. A penstock could be any number of devices, but is most commonly seen as a gate. Many systems have enclosed pipe penstocks, which send water to sewerage or hydraulic turbines.

Despite the various methods of categorizing hydropower systems, they all function in a relatively similar fashion. First, water is passed through the system's turbine. Different variables such as head, pressure, and flow rate, all of which can be tweaked dependent on the category of head a system uses, make up potential energy for the system. Then, the turbine converts the water's energy into electricity. The beauty of a hydropower system is that their energy conversion process generally involves little fuss or bother. There is no combustion, no waste, no heat, and, perhaps best of all, no noise. Of course, nothing is perfect, and hydropower comes with its own set of problems, some of them quite disastrous.

4.6: HYDROPOWER APPLICATIONS

Hydropower is most commonly used to generate electricity, but because it is water, its versatility allows it to be applied to many things. Water's kinetic energy, in addition to the energy of which it is capable at a great height, can be directly applied to powering machines. In fact, such was the sole use of water power until the middle of the 19th century.

As a rather stark contrast, the direct use of a water source's energy has dropped considerably in modern times. Old-fashioned water mills can still be found in some locations, usually put to use sawing wood or grinding corn into bits. Counter-balancing tanks fill their tops with water, which is emptied at the foot of a hill at some mountain railways. Again, though, such instances are considered extremely rare in our modern, industrialized world.

Not all places are so industrious. Many developing countries depend upon water power for many tasks. Nepal features simplistic turbines that are locally produced and subsequently used to power machinery. In Asia and various Middle East locations, flowing streams raise water for irrigation, just like the ancient noria used to do.

4.7: THE FRANCIS TURBINE

As mentioned earlier, James Francis is the creator of the Francis turbine, the spiritual successor to Uriah A. Boyden's Boyden turbine. The Francis turbine is by far the most common turbine found in the modern world's medium- to large-scale plants, and is primarily used in installations involving a small head of approximately two meters, or high heads as great as 300 meters.

You may recall that the Francis turbine was and is popular due to its ability to run while completely submerged in water. Additionally, the Francis turbine is perfectly compatible with both vertical- and horizontal-axis system configurations. Its adaptability is what made and what has kept the Francis turbine a mainstay in modern water power systems.

During use in high- or medium-level heads, the turbine channels water through a volute, a scroll-shaped tube that diminishes in size while wrapping around, much like a snail shell. Due to the snail-like design, the volute forces water to run toward the runner. In such a system, the guide vanes will be set in the turbine's inner surface, assisting the water in its path toward the runner.

As water crosses the curved runner blades, it is deflected sideways. Besides causing the water to lose its whirling motion, the deflection forces it in the direction of the axis, where it is then sent out along the central draft tube, and then to the tail race. The effect of the water being deflected is its force, which pushes the runner blades in the direction in which they are traveling, sending energy to the runner and ultimately keeping the rotation going strong.

Many factors come into play, all of which result in the Francis turbine's high level of efficiency. The shape of the guide vanes, in addition to the runner blades, combines with the water's speed to produce a nice, smooth flow. Instrumental to a high efficiency rating is the speed of the runner blades. If greater or equal to the speed of the water, efficiency ratings will fall. Optimally, runner blade speeds should be slightly less than the speed of the water.

The Francis turbine is not without limitations. If the water's head is too small, a dramatic volume flow will be necessary for a given power. Also, a low head can often equate to a low water speed. Both aforementioned factors require a larger input area.

4.8: EFFECTS ON ECOSYSTEMS

As powerful as water is, it comes with a rather large set of problems. The installation of a water power system will usually result in serious repercussions on the given area's ecosystem. The surface water, for example, will be come warmer, which will lower its oxygen content. This will not bother some species, but it can harm others. For example, smallmouth bass will thrive due to this dramatic change, but salmon will drop off considerably.

Dramatic fluctuations in a system's water level will most likely cause difficulties for wildlife, both animals and plants, dependent on water found near the area's shoreline. Various fluctuations will cause the water level to decrease, which will have an immediate negative effect on the wildlife dependent upon the shoreline for food and cover.

Also, sediment will invariably build up in the reservoir, which will result in modifications in both the reservoir's immediate area, and the downstream area.

The construction of dams makes fish passage more complicated than it should be for such simple creatures. The construction of fish ladders, structures such as steps which serve to aid fish in their transportation, provides a bit of help. The trick to building a useful fish ladder is to make sure that the velocity of the water falling over a step setup is great enough to attract the fish to the ladder. However, the water's velocity cannot be so great as to wash the fish back downriver, or even worse, exhaust them to the point that they are unable to continue their journey up the river.

4.9: ENVIRONMENTAL ISSUES

Water power systems come with some adverse consequences, but they have their fair share of positive environmental effects as well.

Most beneficially, water power systems release no carbon dioxide (CO_2) into the environment. They do release small amounts of sulphur and nitrogen, but they are so small as to be negligible, which is good, considering those two elements are heavily involved in acid rain.

On the other hand, land flooded for a hydro-specific use can emit very high amounts of methane gas, which is a far more potent gas than CO_2. Armed with that knowledge, it may seem proper to classify hydro-based electricity generation as large a contributor to global warming as fossil fuels.

However, contrary to some fossil fuel usage, and even some renewable energy sources, water power systems are not radioactive, nor can they bring about fires or explosions. In those respects, hydropower systems are quite safe.

Unfortunately, hydropower systems can be quite unsafe in other, vastly important ways.

If a dam collapses, tens of thousands of people can lose their homes, livelihoods, and/or lives in a heartbeat. In May of 1889, the Johnstown Flood, known as "the Great Flood of 1889," occurred when the South Fork Dam, located approximately 14 miles upstream of Johnstown, Pennsylvania, failed due to several days worth of constant rain. Twenty million gallons of water flooded through

Johnstown, vanquishing homes and animals, and killing over 2200 people, including 396 children.

En route to the town, the tidal wave picked up debris ranging from trees, rocks, and mud to railcars and barbed wire from factories. The water slammed into Johnstown at an estimated 40 miles per hour. Not only did people drown, many were crushed and maimed by the debris carried into town via the water.

Though not as physically harmful, the construction of large dams is responsible for thousands of relocations. During the 20th century alone, the construction of massive dams has directly resulted in thousands of people being forced from their homes and property. Building the Kariba and Aswan dams entailed the relocation of 60 thousand and 80 thousand people, respectively.

Despite the negatives, building a dam can have a positive impact on a region. For those who live on a river that constantly overflows, the construction of a hydropower system might mean relative safety from such dire happenings.

RIDING THE WINDS OF CHANGE - WIND POWER

5.1: A HISTORY OF WIND POWER AND WINDMILLS

Despite water power sources being considered the earliest renewable energy source responsible for significantly reducing the workload of man and beast, wind did not lag far behind. Using wind to generate different forms of energy is a technique that dates back thousands of years, though the technology behind this vital renewable energy source has evolved considerably since its inception.

Traversing the annals of history reveals that a variety of civilizations were able to harness the power of the wind in order to make their daily lives easier. Many primitive devices for performing labor-intensive tasks, such as pumping water and grain, have been used so long that it is almost impossible to trace their roots.

A common assumption often made regarding wind power is that "windmills" are used to produce the energy that is eventually converted to electricity. In truth, not all wind power devices are called "windmills." Windmills have been in use for over 4000 years, and are actually used for milling grain.

The devices most people typically mistake for windmills are known as "wind turbines," and are named as such due to their functionality, which is very similar to both gas and steam turbines. Additionally, the title of "turbine" is also important in distinguishing them from their windmill predecessors.

Given that windmill history stretches back over four thousand years, it should seem obvious that the design and functioning of the windmill has undergone countless reiterations.

- The "screened" windmill features either screens or walls, and sometimes both. The walls are constructed of a similar material to screens, since airflow is necessary in the design of a windmill, and they are positioned to screen the windmill's sails from wind during the reverse part of the wind cycle (when the device is moving into the wind).

- The "Clapper" windmill makes use of moveable sails that "clap" against their stoppers. The Clapper's rotor is rotated forward via the power of the wind, which serves to maximize wind resistance. During the reverse phase of the wind cycle (again, the time during which the device is moving into the wind) the device is altered to allow a reduction in wind resistance to take place.

- The "cyclically pivoting sail" windmill is similar to the Clapper's design, but is comprised of far more complex mechanisms that boast dramatic changes in sail orientation during the windmill's rotation cycle. During the rotation cycle, the angle of every sail is slightly adjusted according to not only its position, but the direction of the wind as well. Such an adjustment to each sail results in different levels of wind resistance on each side of the windmill's rotation axis, which will rotate only when exposed to a healthy, constant flow of wind.

- The "differential resistance" windmill might also be referred to as the "differential cup" windmill. This particular version of the device sports blades that are specifically shaped to offer greater resistance to the wind on one surface when compared to any other. Adhering to this particular design produces a different degree of wind resistance on each side of the axis, and subsequently allows the windmill to turn.

Each of the different windmills discussed shares a common trait: they are all vertical-axis windmills, which were popular early in the windmill's history. Later, we will be studying vertical-axis windmills in depth, in addition to an equally close study of another breed of windmill: the horizontal-axis design. For now, a brief overview of each design will suffice. For those readers interested in skipping ahead, Section 5.3.1, "Horizontal-axis Turbines (HAWTs)," and Section 5.3.2, "Vertical-axis Turbines (VAWTs)," can be read for more about each of these fascinating technologies.

Horizontal-axis windmills originated sometime during the 12th century, and were most commonly found in Europe. Most early versions of the horizontal-axis windmill were comprised of pistons that churned in and out from a central location. This radial design is accompanied by different sails that are dependent on the arms for support. The sails rotate around the axis in a plane that faces the direction in which the wind blows. Each sail and blade was specially angled toward the wind and moved at right angles in the wind's direction.

After their use grew more popular, it was not uncommon to see two main windmill designs all throughout Europe. The "post mill" featured a windmill body that was moved along a large, upright post. Whenever wind fluctuation occurred, the post mill's movement took place immediately afterward. The second windmill design, known as the "tower mill," was much more popular, and much more common. This particular design was outfitted with a rotor and a cap, both of which were supported by a very tall tower. The cap would move whenever the wind direction underwent a change, and the sails in turn would move slowly, thereby generating mechanical power.

Before the era of the Industrial Revolution ushered in fossil fuel technologies, it was not considered unusual to see more than 10,000 of Europe's two most common windmill designs scattered here and there throughout Britain.

As is the case with any renewable energy technology, wind power has been through countless refinements, with each new generation striving to reduce wind power's costs and achieve maximum efficiency. As of 2002, more than 31,000 megawatts (MW) of power were being generated by windmills. That astonishing figure is approximately four times the amount of wind-generated electricity than had been installed by the end of 1997. Such growth resulted in a roughly 40 percent increase every year in between.

Though commonly thought of as powering large turbines, the power of the wind can be used for smaller applications as well. Since the 1990s, many small battery chargers, fully powered by wind, have been in use. Though their efficiency has not been directly compared to solar battery chargers, each is an efficient, environment-friendly method of recharging a technology that is ubiquitous in home electronic devices.

Battery chargers powered by the wind are only one of the wind power marvels ushered in by the '90s. Perhaps the most important in relation to wind turbine efficiency was a dramatic increase in the possible total capacity of megawatts. During the middle of the '90s, most turbines featured a capacity of approximately 500 kilowatts (kW), but during the year 2001, that amount had increased to greater than one MW.

Today, many wind turbine prototypes come equipped with total capacities more than double, triple, and quadruple today's de facto standard of two MW. Even with such dramatic advances, the available power of wind energy remains, and will probably always remain, untapped.

5.2: WIND TURBINES

Before we delve into the specifics of how wind turbines function, it is important to attain a clearer understanding of each of the two different types of turbines: the horizontal-axis turbine, and the vertical-axis turbine.

5.2.1: Horizontal-axis Turbines (HAWTs)

Horizontal-axis turbines (HAWTs) are perhaps the most commonly seen turbines in the U.S. HAWTs have evolved from traditional windmill designs, and they are usually used to generate large supplies of electricity.

Horizontal-axis turbines typically have two or three blades, though some designs feature more. Horizontal-axis turbines with several blades feature a disc that is covered with blades stacking out of any available space; think of the pictures of the sun you probably drew when you were a child. Due to this common design, many horizontal-axis turbines are often compared to aircraft propellers.

Horizontal-axis turbines are also referred to as "axial-flow" turbines, due to their rotation axes being perfectly in line and coordinated with the direction of the wind. Each rotation axis is able to be kept in perfect harmony with the direction of the wind by using yawing mechanisms, which quickly realign the turbine's rotor whenever the wind's direction changes. Because the direction of the wind can change frequently, the yawing mechanism is kept busy by constantly re-adjusting the turbine's rotor.

The performance of all axial-flow turbines is dependent upon several factors. First, the number of blades, in addition to the shape of each blade, is instrumental in properly flowing with and against the power of the wind. Second, but of equal importance, is the choice of the turbine's aerofoil section. An aerofoil is a device that, when in motion, is able to provide the turbine with a reactive force that is relative to the air surrounding the turbine. Third, certain attributes of the blades, such as chord length, the "angle of attack" positions along the blade, and the angle (remember, blades are set to "ride" the wind), are integral. Finally, the amount of allowable twist between the hub and tip of a blade can result in lowered efficiency, if not properly designed.

5.2.2: Vertical Axis Turbines (VAWTs)

Though horizontal-axis ("axial-flow") turbines are far more prolific in their usage, vertical-axis turbines (VAWTs) could be considered much more efficient. VAWTs have the distinct advantage in that they are able to harness the wind completely independently of the direction in which it is blowing, without having to reposition the turbine's rotor due to a change in wind direction. Some might say that this alone makes vertical-axis turbines better than axial-flow turbines.

The man largely responsible for the design of the modern day vertical-axis turbine is Georges Darrieus, a French engineer who designed and produced the Darrieus vertical-axis turbine in 1925. Darrieus' design made great use of large, curved blades, the ends of which were attached to the top and bottom of a long, vertical shaft. Though it has been in use since the early 1900s, the Darrieus vertical-axis turbine is still considered the most advanced of all modern VAWTs.

Another modern type of VAWT, the H-VAWT, is designed as a tower capped by a hub. Two horizontal cross arms attach to the hub itself, giving the H-VAWT the appearance of the letter "H" being raised by a long pole.

Another type of VAWT is the V-VAWT, and its design entails several perfectly straight aerofoil blades that are connected at one end to a hub on the turbine's vertical shaft. The aerofoil blades are inclined in the shape of a "V." V-VAWT towers are typically rather short, as are their ground generators and blade installations.

Due to boasting a much more significant solidity than H-VAWTs, a rotor for a V-VAWT is always more expensive, not to mention much heavier than their H-VAWT cousins. Because of their higher price and weight, V-VAWTs are not competitive with H-VAWTs in terms of pure economics, but constant research is being conducted with subsequent refinements being produced.

Not to be outdone by the "axial-flow" synonym for HAWTs, VAWTs are also referred to as "cross-flow" devices. The "cross-flow" moniker is because a cross-flow turbine ensures that the direction of the wind is at a right angle to the axis rotation. To paraphrase, cross-flow turbines ensure that the wind flows smoothly along the axis. Interestingly, modern vertical-axis/cross-flow turbines extract the most power from the wind as it crosses their front and rear.

5.2.3: How Do Wind Turbines Work?

Creating electricity with wind turbines has much to do with blades, shafts, and angles. The entire energy conversion process is simplistic, though the components involved can be complex.

In short, steady wind speeds turn a wind turbine's blades,

which then turn the turbine's generator, which is then used to generate electricity. For this simplistic process to start and finish properly, a number of major components are involved.

- First, we have the blades, perhaps the most important aspect of any wind turbine. In today's modern world, most two megawatt turbines have two to three blades, each approximately 130 feet in length. Most blades are constructed using glass-reinforced plastic or a similar type of durable material. Durability is important due to inclement weather and, in some areas, strong winds.

- The second component, known as the nacelle, is nothing more than the housing on top of the turbine's tower. The nacelle houses crucial wind turbine components, such as the generator, which is instrumental in actually generating electricity, and the gearbox, as well as several other components that are responsible for converting rotational energy into electricity.

- A variety of trinkets and gadgets, including foundations, transformers, and roads, make up the balance-of-plants components, all of which play small yet vital roles in ensuring a wind turbine's operability and stability.

- Last but not least, every wind turbine includes a tower, the actual turbine itself. Without a tower, blades, transformers, generators, gearboxes, and everything else will have nothing to attach to, thereby rendering the wind turbine useless. Most modern towers with a capacity of two MW are approximately 250 feet high.

5.3: APPROPRIATE WIND SPEEDS

A common myth is that wind turbines need a strong, steady supply of wind in order to generate a sufficient amount of electricity. This is only half true. A wind turbine will usually start generating electricity upon reaching wind speeds of 12 to 15 miles per hour. Each wind turbine has a rated level of output that is reached when wind speed reaches approximately 30 miles per hour, which is an important line that should not be crossed.

As wind speeds approach 40 miles per hour, electricity will still be generated, but the risk of damage grows. If wind speeds upward of 50 miles per hour are attained, most turbines will shut down in order to avoid damages. Strong but stuttering winds can sometimes produce a sufficient amount of electricity, but overall, it is far better to rely on a steady stream of wind rather than scattered bursts of powerful wind.

Because wind turbines begin to generate electricity at 12 miles per hour, any area of the world featuring steady wind at approximately the aforementioned speeds is considered an ideal site for one or more wind turbines.

5.4: ESTIMATING THE POWER A TURBINE COULD PRODUCE

Estimating the power a wind turbine is capable of producing

is not as easy as simply measuring the electricity it generates. Several factors are integral in the equation of determining electricity potential, one of which is known as the wind speed-power curve. The wind speed-power curve takes into account the area swept by the wind turbine's rotor, the choice of the turbine's aerofoil device, the number of blades attached to the turbines, the shape of each of the blades, the optimal blade tip speed, how fast the turbine will be able to rotate (this factor alone is dependent upon blade shape, size, and angle), the aerodynamic design of the wind turbine's blades, and the efficiency levels of both the gearbox and the generator.

Equally important is the wind speed frequency distribution, which is simply a graph that depicts the number of hours during which the wind blows at a variety of speeds for a set period.

5.5: COSTS OF WIND POWER

Intertwined with the power a wind turbine might produce under ideal conditions is, of course, how much it will cost to accomplish that goal. Because so many components are involved in the make-up of the wind turbine itself, estimating a wind turbine's cost involves many variables.

- The first aspect of the wind turbine we must take under consideration is timing. Whether the turbine is already in use, is currently being installed, or Whether its future installation and use is being planned, timing is vital in gauging a turbine's cost.

- Another consideration is whether the wind turbine

or wind farm will be dealing with the market price of electricity, or the initial cost per kilowatt hour (kWh)?

- Sometimes assumptions are made that need to be converted into definite conclusions. For example, whether discounted rates will be available.

- Care should be taken to gauge the wind resource properly. Determine the lowest, highest, and average speeds. Remember, 12 to 15 miles per hour is what is considered necessary for sufficient electricity generation. Also determine how long the wind will continue to blow once it starts. If the wind is intermittent, the site will end up costing much more due to constant downtime in electricity generation.

- Finally, most vendors tend to vary their prices due to factors such as different types of technology. Those in charge of prospective wind turbine construction should shop around before a final decision on a seller is reached.

A common approach to determining the cost of wind power is to sort out the price per individual component. Everything from turbines and their individual parts, such as blades and gearboxes, all the way to operating costs should be thrown into an equation that results in a profitable per-kilowatt-hour price. Some components, such as the wind itself, require more work. The initial cost of the turbine itself should be counted with operating costs, which entail things such as maintenance and service fees.

Of course, even the most dedicated equation breakers will

come up with what amounts to estimation. In the world of wind power, almost everything fluctuates, such as whether the land on which to build the wind turbine(s) will be purchased, or leased. If the land is purchased, that must be factored into the initial costs of the project; leasing falls under the category of operating costs. The leasing route has a few variables itself, such as whether the lease will be based on an annual rent payment plan, or on capacity (in terms of MW) or output (kWh).

Another factor to consider is initial costs. These can be separated into the cost of the turbine — these costs typically run from $650 to $700 per kilowatt (kW) — and balance-of-plant components, such as foundation, the accessibility of the electrical connection, and the use of the land.

Determining the operating costs themselves is largely based on objects comprised of the output, such as gearbox maintenance and keeping the blades clean, and per year charges, such as rent and insurance.

5.6: SITING IDEAL LOCATIONS FOR WIND TURBINES

Properly siting an ideal location for a wind turbine, or more commonly, a group of wind turbines, known as a wind farm, is simple in the sense that wind blows almost anywhere and everywhere, and in that respect, choosing ideal locations for wind turbines is, as they say, a breeze. A recent study revealed that in the United States alone, 45 of the 50 states possess enough adequate wind sites to justify installing turbines.

The locations with the absolute best wind sites are the Great Plains states, comprised of Colorado, Nebraska, Montana, New Mexico, North Dakota, Kansas, South Dakota, Texas, Oklahoma, and Wyoming, and the Canadian provinces of Alberta, Manitoba, and Saskatchewan. Again, those are just the best sites for the United States and Canada; many more exist worldwide. A review conducted by the European Unions proclaimed that Europe could procure 5 to 10 percent of its total electricity needs solely from wind power. Another more recent study revealed the belief that Europe could probably gain closer to 20 percent of its necessary electricity from wind power.

5.6.1: Common Siting Problems

With so many ideal locations for wind turbines, you might wonder why we do not install turbines everywhere. The answer lies in the aforementioned difficulty of siting wind turbine locations, with the primary objection to such installations stemming from local stubbornness. No matter where a site may be proposed, many local inhabitants simply do not like the idea of their scenery being obstructed by wind turbines, which they consider a form of visual pollution.

It is a reasonable argument, because many areas that offer good conditions for wind turbine installation also boast tremendous scenic views, which raises the question: should we preserve nature's beauty, or make room for electricity conversion?

Wind turbines are also common perpetrators of noise pollution. Most turbine-related noise pollution stems from its mechanical equipment, such as the common clinks and clanks that emanate from the generator and gearbox; such racket is called "mechanical noise." Because all machines tend to make noise, these two components are usually the cause of the problem. A common way to remedy any source of mechanical noise is to upgrade the equipment so that it is able to take advantage of quieter gears. Enclosing the equipment in sound-proof material also causes significant reduction in mechanical noise. Running the turbine at low speed is also beneficial, as is switching over to a direct-drive generator.

Many modern wind turbines are able to function properly without a gearbox. By so doing, these particular types of turbines eliminate a large source of mechanical noise and can sit much closer to buildings.

The second most common cause of noise pollution caused by wind turbines is "aerodynamic noise," which refers to the interaction between the spinning blades cutting through the air current. As the rotor's rotation speed gradually increases, so too will the amount of aerodynamic noise that runs parallel with it. A common remedy is to design the wind turbine to operate at a lower speed of rotation during periods of decreased wind speeds.

Another solution, and one that can work in tandem with any other, is to increase the wind turbine's total number of blades. As each blade cuts through the air, the thickness is reduced for the next blade in line, which diminishes the "swoosh" sound made by the turbine's blades.

Noise pollution is considered a much more sensitive issue

than visual pollution. Many individuals do not mind the sight of wind turbines, but a constant stream of noise that distracts them from sleeping, studying, and other quiet activities is often considered intolerable. Many countries have regulations detailing maximum noise levels that must be strictly adhered to, with some locations allowing only certain levels of noise during specific times of the day.

Another source of noise pollution, though less common than mechanical or aerodynamic noise, is electromagnetic interference. Some televisions receive interference from television wind turbines. Such types of interference are largely determined by the types of material used to construct turbine components such as the blades, and the surface shape of the tower itself. Some blades can be made of materials such as timber, which aid in absorbing the interference, and thus are able to eliminate the problem.

Great consideration should be given to the amount of noise pollution a single turbine or an entire wind farm will produce. If the community is against the installation, the project may never come to fruition.

5.6.2: "No Fly Zone"

There have been many recorded incidents concerning military complaints about wind turbines. Concerns have been voiced by the United Kingdom's Ministry of Defense that claim wind turbines cause interference with certain types of military equipment, such as radars. In addition, the UK Ministry of Defense also has issues with wind turbines interfering with specific training exercises, specifically those that require pilots to fly close to the ground, in which case the planes would be damaged if a wind turbine was clipped.

Though valid concerns, most of these and other military problems have been solved. Various reports and studies conducted by associations such as Natural Power have concluded that, after dealing with minor instances that have been quickly remedied by all parties, wind turbines and the military have reached a ceasefire.

5.6.3: Birds and Other Flying Critters

The least common problem involves birds flying into a wind turbine's blades. Some communities use this as an objection to wind turbines being installed, but for a comparison, accidents involving fossil fuels, such as oil spills, are responsible for a phenomenal number of bird deaths, much more so than the blades of wind turbines.

The Exxon Valdez oil spill that took place in Prince William Sound was responsible for killing over 500,000 birds, not to mention a variety of other forms of life. Those casualties are an estimated 100 times greater than all wind turbine-related bird deaths.

5.6.4: Some Solutions to Siting Problems

In an effort to reduce cases of public opposition, several solutions have been proposed to deal with various complaints centered on wind turbines. Many remedies for noise pollution, such as those discussed in Section 5.6.1 have already been proposed and put into practice, but what of visual pollution?

In many instances, the entire construction of the turbine and its four main components —blades, the nacelle, all the various balance-of-power components, and the tower itself — are being redesigned from the ground up to be more visually compelling, which research indicates will cut down on population opposition. This effort manifests itself with towers being constructed of tapered, tubular steel, which many find more pleasing to the eye, and, as previously mentioned, in the form of upgrades to components such as the generator, gearbox, and the turbine's blades.

5.7: OFFSHORE WIND

Offshore wind sites, such as Cape Cod, provide a drastically larger amount of wind than any land ("onshore") resource available. Because of this, offshore sites are finding an increasing amount of attention given to them from prospective wind turbine constructors, but they come with their own set of problems.

5.7.1: Energy Potential

Various studies have been conducted to determine just how much potential offshore wind resources might offer. The results are promising. One study found that the wind resource available in the United Kingdom is comparable to that region's entire electricity requirement. The offshore wind resource available in Scandinavian countries is so large that many believe it could exceed all of their electricity needs. Similarly, Europe also boasts a tremendous offshore wind resource, and could easily see all of its electricity resources met and surpassed.

5.7.2: Problems with Offshore Wind Sites

Unfortunately, an entire slew of siting and other problems come with the enormous potential of offshore wind turbine sites. Offshore wind sites are not as commonly traversed as onshore sites, and thus, any wind turbines installed on an offshore site could be overlooked, as they would primarily be viewed at a distance. However, just because people rarely traverse certain paths does not mean that those paths are never traversed; most likely, tourists and natives alike will not like having wind turbines in their faces.

A far greater problem is the beauty for which many offshore sites are praised. Sites such as Cape Cod, Copenhagen, and many others are prized for their scenic views. Disrupting such natural beauty by installing wind turbines is considered shameful by many individuals. This problem is considered far more serious in high population areas, such as many European countries, and is not as prevalent in the less densely populated western U.S. Of course, just because it is not as large a problem does not mean the problem does not exist.

Yet another problem is the finances that construction of offshore wind turbines always entails. The foundations of such sites are more expensive, and unorthodox procedures, such as the installation of transmission cables, must be carried out. Also, just as their locations are coveted for their wind power potential, so are they hard to access. It is much easier to conduct maintenance and service repairs on land-based sites, as they are easily accessible. This is not so for many offshore wind power sites.

Decent savings can be had by implementing solutions such as faster turbine blade rotation. The more air a turbine or an entire farm is processing, the more money it is simultaneously earning and saving. Also, mechanical and aerodynamic noises are of little consequence when applied to offshore sites. Cheaper gearboxes that happen to be louder than their onshore counterparts can be purchased, because, due to a lower population, noise is not as large of an issue.

By far the largest savings to be gained by the use of offshore wind turbines is the electricity they will produce. It will be far greater than any onshore solution. Because they are generating a larger supply of electricity, many investors can be convinced of the long-term savings that are manifested through a larger net electricity gain.

5.7.3: Overview of Offshore Wind Sites

Offshore winds are usually stronger than onshore winds, but more importantly, they are a great deal more constant, and this fact could likely assist in solving offshore wind's few but crucial siting issues. While the visual pollution is a valid argument, many would tolerate a few turbines sitting on their favorite bluff or body of water to have a vastly larger supply of usable energy that could save much more money and fuel sources.

5.8: STRENGTHS AND WEAKNESSES OF WIND POWER

Regarding wind power's pros and cons, I am happy to report that despite a number of valid negative points, the good points, while not as many in quantity, greatly outnumber the cons in quality.

As previously discussed, wind power suffers from certain real-estate problems; namely, that not just any location is ideal for the construction of one or more wind turbines. Some areas, such as the U.S.'s Great Plains states, present optimal locations where wind is constant and develops at just the right speeds: not too slow, but not so fast that wind turbines could be damaged. Unfortunately, most areas across the globe experience variable winds, with spurts of gusts and long periods of stillness or winds too slow for turbines to properly make use of. In order to convert wind energy into a useable form, wind must come at a constant, steady rate.

Of course, even in areas with strong winds, the wind will cease to blow at some point, but this will not occur everywhere at exactly the same time. As long as the wind is steady and strong enough, wind turbines can be built in areas with intermittent periods of wind. However, a larger amount than normal will need to be built, and they will need to be spread around a greater distance so that, should the wind decrease in one area but increase in another, at least some turbines will be on hand to properly capture and convert the wind energy into a useable form.

Even if a location boasts just enough steady wind to put turbines to work, the wind cannot be too variable. Otherwise, the converted wind power will be too difficult to store. Most wind energy is converted into electricity, which must be used on demand. Because of this, most wind energy systems strive to produce a slight surplus of electricity over what will actually be required. This method of electricity production cannot be properly relied on, however. If the wind is too variable to begin with, it is doubtful that electricity will always be able to be generated precisely when it is needed. Anything less will upset consumers and investors.

One alternative is to use wind farms as a supplement to conventional fossil fuels that are used to produce energy. Many contend that using wind energy as a backup power source saves just enough fossil fuels to be considered effective. Typically, a conventional power plant functions normally until a high-wind period becomes available. At that point, the power plant allows wind turbines to contribute their energy until the wind dies down. From there, the power plant is fired up to its normal rate of efficiency until another steady flow of wind occurs.

As advancements in weather forecasting occur, more accurate wind forecasts are available, which serve to increase the efficiency ratings of most modern wind turbines.

Even more bothersome to investors, builders, locals, and tourists is that some locations are revered for a variety of reasons, typically historic and scenic. As we have discussed, locals and tourists object to wind farms being built in such locations due to the visual pollution they create. Problematically, most ideal sites in terms of wind speeds and steadiness are also those that purport beautiful views and vacation spots. To offset such complaints, many modern wind turbines have been painted with muted colors, chosen specifically to blend in with their respective surroundings. This strategy is sometimes, but not always, effective in halting complaints.

Offshore sites not only hold the average greatest potential for wind energy, they are also commonly seen as excellent vistas for tourists. In addition to visual pollution complaints, other locational problems manifest themselves in the form of inaccessibility. Because they are offshore, it requires more labor and finances to reach offshore wind turbines in need of maintenance and upkeep. Most offshore sites produce an enormous surplus of energy, but their problems make planning and hopeful construction a time- and money-consuming process.

Similarly, noise pollution is a problem that persists with many modern turbines, regardless of attempts to mute mechanical and whirring sound as much as possible. It is likely that visual and noise pollution will always plague wind energy technology.

Despite the negatives, wind energy puts forth a compelling case. For capturing, storing, and converting energy, modern wind turbines can be designed to hold a minimum of one or two Megawatts (MW). Additional capacity can be added in incremental stages so that, if properly maintained, no wind turbine will become obsolete too quickly.

Even though visual and noise pollutions can be irritating for all parties, it is a fact that wind energy devices have little adverse environmental impact. Visual and noise pollutions do not harm the environment, no harmful chemicals such as carbon dioxide are emitted, and no waste products are byproducts of wind energy. Additionally, wind energy construction is adept at sharing land. Simply install the turbines and related equipment. No further modifications will be necessary, nor will they exist to interfere with wildlife of any sort.

Similar to arrangements of solar PV arrays, animal wildlife are free to graze around the structures, and plant wildlife will grow freely (as long as the land itself is applicable for such uses).

CATCHING A WAVE - THE POTENTIAL OF WAVE POWER

6.1: INTRODUCTION

The 1980s were a decade of decadence during which all manner of colloquialisms rang out in every direction on Southern California beaches.

You might be paddling your way out to sea when suddenly, the shadow of a towering wave overtakes you. You begin to paddle away from the wet and salty beast before struggling to stand and, miraculously, you tame the creature and ride the juggernaut. Your thumb juts into the air as you shout "Scope this!" to your buds — just before the power of the wave sends you tumbling head over heels into the water where you come up, sputtering, amid good-natured jeering as you crawl onto the warm sand of the beach.

Well, all those surfer dudes and beach bums were onto something. The power of a wave is not only useful for throwing savvy surfers this way and that; it holds enormous potential as a renewable energy source as well.

6.2: A HISTORY OF WAVES

You may wonder why wave power wasn't grouped in with Chapter 4 on hydropower. However, wave power deals with waves specifically, as well as the many different states a sea may take in order to produce waves with adequate potential to manifest themselves as electricity and a variety of other useful energies.

Although the 1980s is known as the peak decade for surfing and other wave-oriented sports, many concepts for producing wave energy date back more than 200 years. As with any form of innovation, these concepts were weeded out and sifted through until only the most efficient remained to be improved upon.

Just as the 1980s were considered the peak years for surfing, the 1970s is largely credited as the years during which people began to become interested in the activity. The same is true for wave power: most of today's modern, viable schemes are derivatives of concepts that came about in the 1970s, a time period which saw an increase in environmentalism, as well as the stifling of conservatism concepts that had been in practice since before the 1950s. In large part, all of our modern methods to properly harvest and convert wave energy into something useable have one common improvement: nearly inconsequential environmental drawback. As you will learn later in this chapter, most (but not all) of hydropower's drawbacks are almost nonexistent in the form of pure wave energy.

It makes sense that the areas that hold the most potential for wave energy are those that boast an active sea environment.

If the water is dormant, no one will benefit from its potential. In areas such as coastal islands, which commonly feature extremely expensive conventional energy sources, wave energy is already considered the best alternative, and in many regions, it will soon be the dominant energy source, renewable or otherwise.

One such country that could quite easily rely heavily on wave power is the United Kingdom. In fact, during an energy crisis in 1973, many in the UK took a high interest in the potential of renewable energy, specifically wave energy. The crisis saw a large amount of wave energy devices undergo numerous tests and applications, but the lack of time to solve the problem and a lack of funds resulted in many experiments being put on hold until now.

6.3: HOW WAVE ENERGY WORKS

Compared to the act of balancing on a surfboard while riding a wave, the act of actually creating waves from which wave energy is collected is quite esoteric; in fact, most of the process is not very well understood at all. Thus far, those who study such things have broken down the process into three steps that sound simple, but involve quite a few variables.

- First, the wind starts out by rippling the water. The airflow causes stress on the water's surface, which is visibly seen by way of waves. Of all three steps, this is considered the most general and easiest to explain.

- The second step also deals with the airflow passing

over the surface of the water. Just as a gust of wind can be broken up into different bursts of air traveling at slightly different speeds, so too can the ripples that birth atop the water. Because the air is so turbulent, each patch of water is subjected to different levels of stress and, thusly, varying fluctuations in pressure levels. When the airflow falls into sync with pre-existing waves, even more waves are produced.

- Finally, when the waves eventually reach a specific size, the wind exerts a powerful force on the up-wind face of the wave. This results in more growth (as well as excited cries of "Surf's up!"), rendering the wave even taller and wider.

Because of those three steps, wave energy is sometimes considered a form of stored solar energy, the power of which is typically strong. The sun is relevant here because each burst of wind responsible for sending the water into a churning, wave-building frenzy is derived directly from solar energy.

6.4: TYPES OF WAVES

Just as there is not one type of water nor one type of wind current, neither is there a common, universal type of wave from which we can harness so much renewable energy.

≋ **Storm waves** are waves that are located close to or directly within the area where they were originated.

≋ **Swell waves** are actually storm waves, but those that have traveled far from their areas. The key point:

they have traveled far, but have undergone low levels of energy loss. As long as swell waves have not lost too much of their energy, they can be viable for wave energy devices.

Regardless of the similarities and differences between the types, a few generalities can be understood about all types of waves. First, the size of all waves is dependent upon a few factors: wind speed (the faster the wind blows, the more stress exerted on the body of water), the duration of the wind and the wave (longer is better), and fetch, defined as the distance over which wind energy is transferred into the sea to form waves.

Larger waves tend to contain more energy per individual meter of a large wave's crest length than smaller waves. Because of this, it is important to reiterate that areas in which high wind speeds can be found are the best candidates for wave energy devices.

It is because of location that the United Kingdom is a prime candidate for the use of electricity and other means of power derived from wave energy. The UK is surrounded by stormy bodies of water, and, as we have learned, turbulent air being swept across water results in bigger and better waves: storm waves, specifically. In addition, because almost all the water surrounding the UK is so tumultuous, all of the swell waves it produces can be harvested with little energy loss due to the stormy water perpetuating itself in all directions.

As if its perpetually stormy waters were not enough, the UK is also positioned at the end of the Atlantic Ocean, which is a long stretch of prime fetch that sees the wind most commonly blowing in the direction of the UK, with both storm and swell waves in tow.

6.5: SEA STATES

Just as a State of the Union address details how we are faring as a nation, the sea state is made up of different factors which serve to determine things such as good locations for wave energy devices, the height a wave could potentially reach, and more.

A sea state is comprised of three important variables: the height, the period, and the character of waves on a large body of water. Each wave has its own unique properties: its period (the time it takes the wave to pass a specific point), its height, and its direction. While one wave is usually not nearly enough to make an impact, it is the combination of every wave on the sea that we are able to view when we observe the surface of the sea.

Because it is extremely difficult to accurately measure even a single wave's height and period independently, it is common practice to use a process that begets an average. In general, the total power contained in a single meter of a single wave is equal to the power of the sum of all of its components.

Given that the sea state in any arbitrarily chosen location is different from those adjacent to it, it should not be unexpected that the power of one area's waves varies greatly compared to that of another area's waves. Obviously, data recorded on different dates and times (regarding both time of day and season) will fluctuate wildly. However, if the specific location is subjected to tests ranging over the course of an entire year, the data gathered should be more than adequate to construct a statistical, viewable chart. For instance, the waves spawned in the North Atlantic contain a high average power density. This renders the North Atlantic as a terrific source of wave energy.

In general, all world areas are subjected to variable fluctuations in wind power and speed. However, those areas with consistent fluctuations should be considered more valuable than those with intermittent, weak wind fluxes.

A wave flows in the direction in which the wind blows. In deep water, the wind can be considered the driver of the wave-based car: if nature decides to bluster to the east, so too shall waves head in that direction. Surprisingly, waves can travel an impressive length without losing a significant portion of their power (this is primarily the reason that some consider swell waves to be only slightly less valuable than storm waves).

Of course, not all conditions will be compatible with one another. Some areas see winds sweep in from all compass directions. It is therefore logical to believe that waves can travel in multiple directions as well, sometimes crashing into one another and losing a woeful amount of their potential energy. In the United Kingdom, wind that has crossed the Atlantic is responsible for waves incoming from a southwesterly direction. At the exact same time, it would not be odd to observe waves generated by a certain type of weather (most likely a storm, or even regular wind currents) moving to the north.

6.6: UNDER THE SEA (STATE)

While waves are an impressive spectacle above the surface, there is quite a bit happening below the surface to ensure waves are stocked with their maximum energy potential. As mentioned previously, water is composed of several hundreds of thousands of particles. To understand the basics of how a wave works underneath its proverbial hood, we must understand that waves are comprised of hundreds of thousands of orbiting water particles. Water can often find itself at a standstill, but for a wave to truly be a wave, it is always in motion.

When a wave is nearer to the surface, each of its water particles is similar in height. In this respect, it might be easy to think of waves as trees. A tree flourishes above ground with branches and leaves extending in all directions. A wave, too, has an arcing face with plumes of water that froth, bubble, and reach. Nevertheless, as we look below the surface, a tree's roots extend deep into the ground. The roots are thick and powerful near the surface, but tend to narrow as they worm deeper into the earth. Similar happenings take place with waves: as a wave's water particles extend deeper below the surface of the sea, the water particles' orbits decrease exponentially in size.

As waves travel from the deep sea to calmer, shallower waters, they begin to relinquish their stored energy before eventually running up to the shoreline and flooding innocent sand castles. Understanding this energy loss is critical to understanding the process of harnessing as much energy as possible from a wave. Because waves relinquish an increasing amount of energy as they come closer to a shoreline, the available energy resource from these waves is ultimately affected. It is not uncommon for a wave initially containing a power density of 50 kilometers or more to finish its journey from sea to shore with 20 kilometers or less remaining. The remaining amount of potential energy in a wave is determined by travel factors, such as just how far the wave has come from its point of origination and the roughness and temperament of the seabed.

Besides traveling long distances, there are other ways potential energy can be stolen from a wave. Breaking waves, more commonly known as breakers, lose an enormous amount of their energy potential.

In what could be considered even worse, breakers are often of such power and magnitude that their collision with wave energy converters and other expensive structures results in the equipment being destroyed. Such unfortunate facts should be kept in mind for those who are looking for locations to install wave-energy conversion machines. If a company should decide that a certain location's breaking waves are to be considered inconsequential due to wave energy potential, any machinery to be installed should be designed with economic functionality in mind.

Never install a device in an area with many breakers if you know the device will be expensive to fix should it be damaged. Instead, make sure that the device will function as economically as possible so that, should it need to be repaired or replaced, it will not put you or your company in debt to do so. Additionally, prepare for the worst-case scenario. Any structures installed in a heavy-breaking wave area should be outfitted to withstand the worst damage possible. Spending a little extra is worth the initial costs if repairs and replacements can be avoided.

A change of direction, known as refraction, occurs when a wave reaches shallow waters. Shallow waters have a lower velocity than their deeper, more turbulent brethren do. Refraction is largely determined by the angle at which a wave crest approaches shallow waters. If a wave crest approaches at an angle, one part of the crest will always touch the shallow water before the rest of the wave. The part that has made first contact will always move at a more sluggish speed than the rest of the wave, which serves to change the overall direction of the wave and contributes to a loss of energy. Obviously, the energy would be nearest its maximum potential if all wave particles were traveling in the same direction at the same time.

A study and subsequent knowledge of a sea area's depth contours allows for techniques such as ray tracing to be used. Ray tracing, and similar procedures, allows for areas with high wave concentrations to be identified and thus marked as prime real estate for wave energy.

Those who study wave energy have determined that the best locations for shore-mounted wave energy converters are areas where available shorelines are formed by steep cliffs. The drop from the steep cliff should end in reasonably deep water. Of course, a logical problem exists with this scenario. For most areas around the globe, shorelines are too shallow for shore-mounted wave energy converters to be effectively used.

6.7: CAPTURING A WAVE

Waves usually hold a phenomenal amount of energy that can be converted into something useable, and harnessing a wave's potential energy is accomplished via intercepting waves using a structure that will emit an appropriate reaction to the force of a wave. This is not a simple process, and, as we have discussed, most applicable processes have only been around since the 1970s. Nothing close to perfection has been achieved to date.

6.7.1: The Process of Capturing a Wave

In order to capture and convert wave energy into electricity or another form of usable energy, we must use a central structure that contains at least one active part. This active part will be used to enact relative movement based on the wave force that will be exerted upon the structure. Our main (or "central") structure can be anchored to the seabed itself or, more commonly, to the seashore. However, the active part of the device must move in response to the wave force, not against it.

As with many renewable energy sources, the "right" type of wave energy device should be chosen based on an ideal location; no one type of structure can or should be considered universal. Floating structures are useable, but a stable frame of reference is necessary so that the device's active part is sure to move relative to the main structure. One typical method of accomplishing this objective is to take advantage of inertia. By building the main structure to a size that spans the girth of multiple wave crests, the active part is almost guaranteed to function properly.

Regardless of the type of the structure, its physical size is instrumental in determining the critical performance of the structure. To determine an appropriate size for all types of structures, the volume of water in a wave should be considered. All wave conversion devices require a large enough structure size in order to capture all of a wave's enormous energy potential.

Ultimately, the size of a structure should be determined by its mode of operation. Any structure will require a swept volume (the total volume of air and fuel mixture an engine can draw in during a complete cycle of the engine) equal to several tens of cubic meters per meter of the width of the conversion device. If the swept volume is any less, the device will be severely limited in the total amount of energy available for capture and eventual conversion. However, devices with a small swept volume might still be capable of capturing energy from smaller waves, though such a scenario should not be considered ideal.

As long as the structure is large enough to handle an immense barrage of water, building the structure larger than needed should not pose a problem. Building the structure smaller than is necessary is an invitation for high maintenance costs.

6.7.2: Specific Designs for Wave Energy Converters

Just as there are prime areas where wave energy is considered more substantial than in others, so do such areas exist for conversion devices. Wave energy converters are classified primarily based on location, but they are also classified in terms of how they go about capturing energy from waves, and it could be argued that this later method of classification is even more important. Additionally, wave energy converters can also be classified based on factors such as their geometry and orientation. From which direction will the wave originate, and in which direction will it face when it eventually reaches a wave energy converter? These questions and more have to be answered for an energy converter to properly capture and convert a wave's massive amount of potential energy.

Because waves gather and lose energy at deep sea, near the shoreline, and in transit between these locations, it makes sense to learn that deep, shallow, and intermediary wave energy converters exist to garner energy within their respective locations. Some energy converters are fixed to the seabed in shallow water. Some can be found floating offshore in very deep areas, and others still are tethered in slightly deep, slightly shallow waters, meaning they are not optimal for extreme depths for which others are designed.

As mentioned, some wave energy converters are specifically designed based on the orientation and geometry of the waves coming in to certain areas. One example is known as a "terminator device," which is built so that its principal axis is parallel to the front of the wave. A terminator device is also built to intercept the wave physically, and because of this, it should be specially designed to absorb much physical punishment. An attenuator device, much like the terminator, is also designed based on geometry and wave orientation, but contrary to the build of a terminator, an attenuator device sees its principal axis set perpendicular to the front of incoming waves. This is so the energy churning within the oncoming wave can be gradually drawn toward the device as the wave moves past it.

A point absorber is designed to draw a wave's energy from the water beyond the physical limitations of the device, but they have rather small physical dimensions relative to the length of a wave. Popular due to being available in many different forms, a point absorber could take the shape and size of a slim cylinder, which would carry out its large vertical excursions via a response to incident waves. The hardware involved in constructing a point absorber typically dictates that absorbers will be a few meters in diameter, but they can absorb wave energy from approximately two times their physical width.

All wave energy converters need to be quite large; this is because the larger the wave and the more kinetic energy it is using, the bigger the device would have to be in order to properly capture and convert all of the wave's energy. Some smaller devices exist, but it makes sense to go after larger waves due to their greater capacity for energy.

The necessary size of wave energy converters dictates that it can sometimes be rather difficult to properly model and test prototypes in order to gauge their potential effectiveness. Because of this, most wave-energy converters are initially constructed as models that require a vast amount of space to function properly. In the days of the first computers, they required rooms the size of gymnasiums in order to carry out their tasks, and such is the case with most models of wave energy converters.

6.7.3: Types of Wave Energy Converters

Knowing a bit about the overall design of terminators, point absorbers, attenuators, and other wave energy conversion devices is good, but let us now discuss specific examples of terminators and other such builds.

6.7.3A: Oscillating Water Columns

The vast majority of energy devices that are built and tested are known as oscillating water columns (OWCs). The design of an oscillating water column is built around an air chamber that punctures the surface of the water. Any contained air residing within the air chamber is forced out of and then back into the chamber via any approaching wave crests. During the air's tumultuous journey back and forth between the air chamber, some of the air travels through a turbine generator that produces electricity.

Perhaps the most common type of turbine is the Wells turbine, which is used in several different types of oscillating water columns. Somewhat a recent piece of technology, the Wells turbine was developed during the late 1980s by Professor Alan Wells. During his tenure at the Queen's University of Belfast in Belfast, Northern Ireland, Wells built his machine to rotate in one constant direction, regardless of any airflow heading in or out of the chamber. The Wells Turbine has aerodynamic characteristics considered more than suitable for wave applications; thus its ubiquitous standing in the field of wave energy conversion.

In 1985, one of Norway's leading companies in the field of engineering, Kvaerner Brug, premiered the Norwegian oscillating water column. The proper abbreviation is actually "MOWC," due to the device being equipped with multi-resonant capabilities. The Norwegian MOWC's chamber was placed snugly into a cliff face that dropped straight down to a water depth of 60 meters. The arrangement of the device allowed for the creation of two harbor walls at the entrance to the Norwegian MOWC, which allowed the system to absorb high amounts of wave energy over a diverse range of wave periods.

The Norwegian MOWC relied heavily on a Wells turbine, which rotated at approximately 1500 revolutions per minute (RPM). As though the power of the Wells turbine were not enough, a 600-kilowatt generator was coupled with the turbine. Together, the dynamic duo of wave energy sent output from their coupling to a power grid, but only after it had been properly converted to useable energy. A winning combination of fantastic performance and low-cost electricity manifested itself as a very happy development team, but their joy was short-lived.

In December of 1988, two severe storms tore the Norwegian MOWC's column from its nest, and as of this publication, the system has still not been replaced. However, many man-hours of re-design have been put into the MOWC's resurrection. A point unanimously agreed upon by all involved as being of the utmost importance is a higher durability. As we have discussed, wave energy converters are constantly subjected to physical abuse, not only from the waves from which they harvest energy, but from other nature-based elements as well.

The use of oscillating water columns is popular, but not mandatory. A fascinating example of a diversion from the water-beaten path is the Pendulor, a prototype that has been thoroughly tested in Japan. The Pendulor's components entail the use of a mechanical link between a moving component, such as a hinged flap, and a piece of the device that is fixed firmly in place. The Pendulor itself is a gate which is fitted at about one quarter of a wave's length from the back wall of a caisson, an underwater structure that retains water by sealing it in, but can pump out the water as well.

A Pendulor's gate is located at the first antinode, a point of maximum amplitude from not an individual wave, but an entire series of waves. This arrangement allows the gate to be subjected to any movements made by a wave. A "push-pull," give-and-take system then converts the Pendulor's mechanical energy into useable electrical energy.

6.7.3B: Shore-mounted Technology

For shallower waters, it is ideal to use machines known as shore-mounted devices. In all of its many forms, this technology is specifically engineered to capture energy from waves that are typically smaller in size and speed, given that they have made it all the way to the shoreline. While they will generate less overall power than their deep-sea brethren will, many consider shore-mounted devices to be somewhat more concentrated. Their primary advantage is their closer proximity to power grids. Given that deep-sea wave energy converters are far from the shore, more work must be done in order to transfer and convert wave energy. That process will inevitably result in some lost energy potential, but shore-mounted devices work with water that has already endured traveling long distances.

Another equally important advantage to a shore-mounted device's distance when compared to a deep-sea device is that of its maintenance. Instead of putting together a crew and traveling far distances to repair deep-sea technology, a shore-mounted device is much easier to reach. This is pivotal to the device's longevity, due to storm waves often crashing against the shoreline and potentially ruining any unprotected equipment.

Unfortunately, just as distance is a great boon to shore-mounted technology, it is also a serious weakness. Locations with weak shoreline waves are not suited to shore-mounted devices due to their minimal wave energy potential. For the most optimal output, the device should be positioned in a small tidal range area, otherwise performance can, and most likely will, be compromised. To make matters even more expensive, each location has different requirements that must be met accordingly. Much like wind and hydropower, not just any area is good for wave energy conversion.

6.7.3C: Floating Devices

One popular technology ideal for catching and converting wave energy is a floating device, such as the Whale. Simply put, floating devices are able to more easily deal with the great wave power density found in offshore locations. Additionally, because the devices can simply be tethered to float idly atop the surface of the water, there is not as much restriction when it comes to their placement, and large arrays of floating devices can be constructed all across the open seas.

Fittingly, the aforementioned Whale structure is epic in size. Weighing in at over 1,000 tons and measuring at 50 meters long, its enormous build is necessary in order to ensure as stable a frame of reference as possible. The Whale has been tested at both a full- and model-scale, the latter of which is not much smaller than its proper girth and length. Thus far, Whale research teams have found the device to be cost effective due to its ability to not only generate electricity, but because it can also cater to leisure activities (such as breaker wave-based sports) as well.

Figuratively swimming alongside his larger sea buddy is the Clam. Developed at Coventry University in Coventry, United Kingdom, during the 1980s, the Clam uses 12 interconnected air chambers with cells that are attached and arranged around the circumference of a doughnut-shaped object known as a "toroid." Each cell is a Wells turbine and is sealed against the onslaught of the sea by a reinforced rubber membrane. As waves sway back and forth, air moves between each of the Clam's attached cells.

As the air moves from one cell due to the impact from an incident wave, it passes through one of the other 12 cells on its way, filling the other cells as the air system is sealed. The airflow is then reversed as a wave crest changes its position. Typically, the clam is deployed in water as shallow as 40 meters or as deep as 100 meters.

Breaking the trend of sea creatures is the Floating Wave

Power Vessel (FWPV). An FWPV is designed with the intent of producing a maximum power output of 1.5 megawatts of electricity, which is equal to 5.2 million kilowatt hours of electricity per year. During its operation, an FWPV captures waves that run up the device's front face, which is a slope. The captive water is then returned to the sea via a Kaplan hydroelectric turbine, a device developed by Viktor Kaplan that features a propeller with adjustable blades.

Professor Stephen Salter is credited with conceiving the idea for the Edinburgh Duck at the United Kingdom's Edinburgh University during the 1970s, the decade during which most wave-energy concepts were born. The Edinburgh Duck features a spine that is carefully oriented as closely as possible to a principal wave's front. This particular makeup classifies the Edinburgh Duck as a terminator wave-energy converter.

The Edinburgh Duck's spine is oriented in its particular way in order to closely match the orbiting motions of a wave's water particles. This is extremely difficult to accomplish, as the orbital motions of one particular wave can be close to perfect at one frequency, yet fluctuating wildly at another. If a long wave can meet efficiency expectations, that efficiency can be further improved by way of the spine of the Duck being flexed via its joints.

In theory, the Edinburgh Duck is one of the most efficient of all wave-energy conversion schemes. Unfortunately, taking the concept from paper to actual execution is vastly more difficult and time-consuming. A full year or more is necessary in order to fully develop the proper engineering which is necessary to construct the Edinburgh Duck at full-scale. In addition, the body of the Duck is constantly rocked by waves. While this does not pose any concerns of damage — the body is very durable — it does make extracting wave energy from a constantly moving body more difficult than most would like.

In an odd case of evolutionary hierarchy, a slightly altered Edinburgh Duck concept can be credited with producing the Pelamis, or "sea snake" wave energy converter. This serpentine energy conversion device is comprised of a number of cylindrical sections that have been hinged together. The arrangement classifies the Pelamis as an attenuator device because each particular segment is an active, moving device.

As the waves rock the sea snake back and forth, the cylinders respond to the motions but are firmly resisted at joints by hydraulic rams. The rams are responsible for "ramming" high pressure oil through the sea snake's hydraulic motors. This act of pumping is accomplished using accumulators that smooth the oil as it makes its journey. Eventually, hydraulic motors power electrical generators that are responsible for a wonderful result: the production of electricity.

Given the inherent flexibility and maneuverability of snakes, it is no surprise that the Pelamis device itself is just as robust in operation. Because the sea snake's spine is not subjected to full structural loadings that would most likely be imposed on it in the event of a storm, the sea snake is said to be capable of inherent load shedding. Pelamis is also able to slither up and down waves rather than position itself perpendicularly across them. This enables it to become detuned in storms, where long waves can sometimes be greater in size than the snake itself.

6.7.3D: Tethered Devices

A diverse change from slithering sea snakes, paddling ducks, and roaming whales, tethered devices are kept on a leash. They are anchored to the floor of the seabed and contain a visible structure that floats on the surface of the sea. Though they are not necessarily visually obtrusive, the visible structure of a tethered device often attracts attention.

Tethered devices fall into the classification of point absorbers, and act as such by way of drawing in wave energy from a far greater width than their physical breadth seemingly allows. Along those same lines, the actual capture width of a tethered device is slightly less because of their limitations of the vertical amplitude, which relates to the motion of the absorber.

Tethered devices make use of latching, which entails holding the floating piece of the device under water for just about one full second before allowing the device to follow the wave. The process of latching was developed to maximize the amount of captured wave energy by allowing the floating piece large amplitude of motion. This is necessary for optimum performance.

The Hose Pump Wave Energy Converter is an excellent example of an efficient tethered conversion device. It had better be efficient, given that it has been in development for more than twenty years at Technocean in Sweden. The Hose Pump Wave Energy Converter uses a hose pump with a reinforced vertical cylinder made of rubber, which is anchored to the seabed at one end and attached to a floating device at the other. Cords run through the tube, and by winding these cords as spirals, the seawater moving through the tube will be pressurized when the tube is stretched via the floating portion of the conversion device. This occurs as the floating device moves upward in response to a wave crest: as the wave rises, so too does the floating device. Seawater is eventually pumped out of the tube and up into a land-based storage reservoir.

Another prime example of a tethered device is the Interproject Service Converter, or "IPS Converter." Conceptualized and built in the 1980s at the Interproject Service AB in Sweden, the IPS Converter makes use of a long buoy with a tube open at both ends; the tube is attached underneath the IPS Converter's main structure. A piston inside the tube is linked to the buoy, and the interaction of the buoy with the water that is caught within the tube is directly responsible for the extraction of wave energy.

The IPS Converter has undergone a few overhauls in recent years. One newer configuration, known as the Sloped IPS Converter, operates at an angle to find additional energy to capture and subsequently convert. The Sloped IPS Converter also makes use of a hydraulic accumulator to provide smoother output, as well as offer the exciting prospect of a firmer power for smaller electricity networks, such as those found on small islands.

6.7.3E: Research and Development in Wave Energy Conversion

Just as it has since the 1970s, research and development in the field of wave energy conversion is constantly moving forward as researchers and potential investors eagerly search for the next big thing in wave technology. While 100 percent efficiency will most likely always be elusive, getting as close to perfect as possible is a constant goal.

Many wave energy researchers support the McCabe Wave Pump as an efficient means of energy conversion. The McCabe pump uses three narrow steel pontoons, which are hinged together across a beam that has been positioned directly into oncoming waves. Each of the steel pontoons moves relative to the position of the others, and wave energy is extracted from the pontoons' motion using hydraulic rams that have been mounted between the hinges of all three pontoons. The result is a form of energy that can be used toward applications such as water desalination and electricity generation.

Other groups are in support of the Wave Energy Converter, developed by Ocean Power Technology, a leading renewable energy company based in the United Kingdom. The Wave Energy Converter resembles an oscillating water column (OWC) in design but comes equipped with a float that is situated on top of the internal water surface. The bobbing and shifting motions of the float are used as hydraulic power, which drives a turbine and eventually generates electricity.

There are many other devices in development, all of which boast high efficiency ratings and low costs, which certainly appeal to investors. Wave energy conversion technology is expanding exponentially, and that can only have a positive outcome due to our inevitable dependency on renewable energy sources.

6.8: INTEGRATING WAVE ENERGY INTO OUR EVERYDAY LIVES

Let us now discuss ways you can use all of that hard-earned, newly converted energy.

First, that converted energy is not some intangible ball of power that has to be applied to something specific, such as electricity. The energy can be used directly, but such a topic is beyond the scope of this book. More likely, converted wave energy will be fed into an energy grid and used from there. Typically, the aforementioned grid will be rather small and used to supply power to a small, remote community. A word of caution: extreme care must be taken when integrating energy from a wave energy scheme. Such energy tends to fluctuate wildly, which will in turn cause dangerous swings in voltage of the grid's frequency levels. Proper precautions such as grounding should be taken by engineers.

One way to smooth out the wrinkles that result in feeding converted wave energy into a grid is by adding other wave energy units to the grid. The total power output will be much smoother than that of a single-wave energy source, and any wild fluctuations in output will be less significant if the energy is delivered to a much larger grid than those commonly used in smaller communities. Areas such as the United Kingdom use massive power grids that are strong enough to absorb any fluctuations that occur, and thus are prime candidates for the use of multiple converted wave energy sources.

Another important factor to keep in mind is that wave energy varies from location to location. Integrating converted wave energy is a much simpler process if the location where the energy is to be used features the proper climate and wave speed, size, and frequency of occurrence.

6.9: THE EXTREME ECONOMICS OF WAVE ENERGY

As wave energy capture and conversion technologies continue to improve, one noticeable facet of devices such as the Whale, the Clam, the Duck, and the Norwegian OWC continues to grow: cost. Just like renewable energies such as PV arrays, wave energy converters are sometimes cost prohibitive, a factor that does not scare off many researchers and engineers, but does give potential investors something to consider.

Many proverbial corners could be cut in terms of wave energy cost if the areas of operation and maintenance costs could be remedied. The energy contained within powerful waves constantly wrecks havoc on even the most durable of structures, something that minimizes the overall efficiency of what would otherwise be money-saving technologies and processes. Though shore-based wave energy converters are more convenient to repair due to their closer proximity to land when compared to deep-sea structures, repair costs still add up quickly. Though it is true that the sturdiest structures will break down and need a few repairs at some point, many wave-energy conversion devices need more attention due to a steady stream (pardon the pun) of battering, powerful waves.

The benefit of repair costs is that they usually do not rear their ugly, broken heads until a wave energy device is up and running. Sadly, this is not always easy to accomplish, as initial costs are another expensive factor that has kept wave energy from becoming more prolific in the world of renewable energy. The initial (or capital) costs of a wave energy converter, regardless of whether it is shore or deep-sea based, are approximately two times the initial costs of any conventional, fossil fuel-based station.

There is a foil to overcoming the initial costs of wave energy converters, though it is rather difficult. If the running costs of the wave energy station are less expensive than those of a fossil fuel-oriented station, some savings will be gained that can be seen as competitive when compared to some fossil fuel technologies. Of course, many other efficiency-lowering factors come into play once the station is powered and working. Due to waves of varying sizes and speeds representing the byproduct of different sea climates, the load factor of a wave energy station will often be less than a fossil fuel station with a comparable workload.

Wave energy researchers can overcome these serious economic detractors by working hard to ensure that their wave energy conversion schemes are reliable in their energy conversion methods. If the energy from a wave can be converted with as little effort as possible, competitiveness with fossil fuels can be achieved.

As for attempting to eliminate repair and maintenance costs, researchers and engineers should strive to make their structures durable and robust so that they can survive for many years. One very effective way to achieve this is to equip devices with as few moving parts as possible. Through useful techniques, such as extensive prototyping and testing through models, wave-energy device development teams should be able to learn the best strategies for reducing operation costs.

As discussed, everything will cease to work at some point, but the longer lifetime a device has, the more money can be saved — and that makes investors happy. In addition, savings in maintenance, upkeep, and energy conversion are typically extended to us, the consumers, who could potentially enjoy using our modern conveniences at a much more attractive price.

Cutting back on typically high initial costs is attainable by way of searching out good locations. Just as with wind and hydropower, some locations are more ideally suited for wave energy conversion than others. The stronger the wave, the more useable energy to be shared with everyone. Along those same lines, a good location should also be tested (usually via a model of the wave energy device being considered) to make certain that efficiency ratings will be as high as possible. Also, investors must factor in initial costs, maintenance and repair, conversion methods, and any other variables that may arise, such as public concern over possible visual pollution and whether the location has a good fetch. Many of the devices we discussed in previous sections have efficiency ratings of approximately 30 percent, but we should always aim higher.

6.10: LIMITING ENVIRONMENTAL IMPACT

Despite a slew of economic difficulties, one area where wave energy never fails to impress is in its satisfying lack of environmental impacts. Wave energy is responsible for an almost complete lack of chemical pollution. Some devices such as the Pelamis ("sea snake"), which make use of hydraulics, might contain very slight amounts of hydraulic oil, but extra precautions are taken to make sure that these types of substances are sealed within the machine, carefully isolated from the environment.

Contrary to the admittedly miniscule amounts of visual pollution caused by other renewable energy equipment, such as wind farms and clusters of photovoltaic arrays, wave energy devices are not known for being visually unappealing. Some types are more noticeable than others, though. Shore-mounted wave energy converters, for example, are rather obvious and could pose a problem for both tourists and locals who take pride in particular beaches.

It is necessary to weigh the pros and cons of such a scenario. As with wind farms, equipment that could make our energy cheaper and allow us to more frequently and easily enjoy modern conveniences seems to outweigh the negatives of noticing a floating, tethered, or other type of wave energy device.

Also similar to wind farms is the noise pollution that wave-energy conversion devices cause. However, just as many offshore wind farms all but negate this factor due to offshore sites usually being far enough away from human ears, wave energy practically eliminates its noise pollution due to its most essential ingredient: the waves themselves. The sounds of waves crashing against the shoreline and sighing as they are pulled back into the sea is not only considered calming by many, but also serves to drown out most of the noise made by wave-energy conversion devices.

Another positive point is that wave energy devices are typically built to make sure any wandering fish are able to continue on their way by being constructed to ensure that fish cannot become trapped within their parts. The fact that wave energy devices do little to interfere with the migratory habits of fish and other sea life does a great deal to garner the attention and support of many environmentalists.

On the other hand is shore-mounted devices' habit of sometimes interfering with shipping and receiving of various imports and exports. The hazards introduced by wave energy to shipping vehicles may be insignificant, but it has been known to cause certain traffic problems as ships wait to maneuver around larger structures.

Somewhat surprisingly, floating structures do not interfere with processes such as shipping as often as shore-mounted devices (and as we discussed, they do not interfere all that much either) due to being incapable of extracting more than a tiny fraction of storm energy. Also, large structures are often removed once they have reached the end of their lifetime, freeing up more space for ship travel, sea life, or a new device.

Finally, the emission of dangerous carbon dioxide and

other chemicals from wave energy schemes is extremely low, especially when compared to conventional fossil fuels such as gas and coal. Even better, the emissions from wave energy can contribute to goals such as meeting a specific climate change and hitting targeted acid rain projections.

ROCK 'N' ROLL - HARVESTING GEOTHERMAL ENERGY

7.1: AN OVERVIEW OF GEOTHERMAL ENERGY

The concept of geothermal energy applies primarily to heat that comes from within the Earth. While the core of the Earth is hot enough on its own, approximately 45 to 90 percent of the heat used in geothermal processes is believed to originate from the decay of radioactive elements within the mantle of the Earth. Such heat is the lifeblood of an efficient geothermal energy-powered service, such as electricity generators or heating systems.

Another advantage that geothermal energy holds over wind power is its far more simplistic siting requirements. Unlike wind's tendency to be intermittent and therefore only a viable solution to fossil fuels in certain locations, geothermal energy can be harnessed anywhere hot temperatures exist near the Earth's surface.

7.2: SITING PROSPECTIVE GEOTHERMAL ENERGY LOCATIONS

Wind power's title of most intermittent renewable energy

source remains unchallenged. Despite wind power's nearly uncontested championship reign, geothermal energy is almost as finicky; so finicky, in fact, that labeling geothermal as a "renewable" energy source is considered ambiguous at best, or a complete misnomer at worst.

The potential of geothermal energy in any given location is completely dependent upon the location. If temperatures of at least 100 degrees Celsius exist, then the site is considered good enough, but not completely ideal. Conversion of geothermal energy into a usable form is entirely dependent upon the heat of the Earth's surface. Some places boast phenomenal temperatures, some do not. Sites that provide a minimum surface temperature of 100 degrees Celsius are quite rare, and ideally, a location will carry surface temperatures of greater than 150 degrees Celsius.

Other problems arise even after a prime location for geothermal energy conversion has been found. As you will learn in forthcoming sections, geothermal energy is extracted primarily by way of drilling into the Earth. Such practices are expensive, resulting in geothermal energy prices that are almost comparable to PV technology.

7.3: GEOTHERMAL PLANTS

Compared to the typical 25 percent efficiency ratings generated from wind farms, geothermal plants operate at an impressive average efficiency of 90 percent. This high efficiency rating ensures that geothermal energy is a reliable source of power in locations where it can be used.

Most modern plants make use of either steam of binary

cycles to convert geothermal energy into a usable form, such as electricity. A binary cycle consists of mixing water with various other minerals that have been pumped from a geothermal well. The heat of this mixture, known as "brine," is then extracted with a heat exchanger. The cool brine is injected back into the Earth, where it is ideally recycled to be used in a future binary cycling process. Binary cycles are most commonly used when dealing with water temperatures in the range of 100 to 150 degrees Celsius. The binary cycling process is much more expensive and complicated than the steaming process, which is one reason that temperatures greater than 150 degrees Celsius are considered ideal.

Steaming procedures entail hot water being "flashed" to steam by reducing the water's ambient temperature. The resulting steam is then used to directly drive a turbine. This procedure is obviously less complicated in its description than a binary cycle, but is also even simpler in execution: if the water has already been converted to steam due to extreme heat, the middle process can be overlooked and the steam put directly to work in driving a turbine. In reality, flashing water into steam is the most common process.

A more modern process, known as "hot dry rock," has been put to effective use in certain geothermal energy plants. With hot dry rock, the heat stored in deep rock formations that lack steam and water is unearthed via drilling equipment. Hot dry rock is an effective process, but it is also one that comes with severe technical limitations. Drilling into deep rock is expensive and time-consuming, as is transferring the resulting fluid — most likely water — and fracturing the rocks to allow fluid penetration. The potential of hot dry rock is massive, but the costs of drilling and the other aforementioned limitations have kept the process from becoming ubiquitous with geothermal energy.

7.4: USES FOR GEOTHERMAL ENERGY

For locations that can afford to extract and convert geothermal energy, as well as produce temperatures hot enough to meet minimum geothermal requirements, geothermal energy has a multitude of uses. Direct-use geothermal is the process of using hot water or steam immediately after extraction. The water and steam are commonly used for space heating applications and other industrial processes. Geothermal electricity sees the water or steam used toward the production of electricity, and is perhaps the more frequent type of geothermal process in use today.

For most places, high prices and fussy siting requirements are worth the end result of geothermal energy. Remember, we have learned that the efficiency rating of geothermal energy is, at worst, 45 percent, and at best, 90 percent. Such statistics present a compelling argument for the perpetuation of geothermal technologies, but therein lies a critical problem.

7.5: PROBLEMS WITH GEOTHERMAL ENERGY

As mentioned, a crucial flaw in geothermal energy lies in its semi-intermittence; not every area on Earth will present prime conditions for the extraction and conversion of geothermal energy. However, a more serious problem manifests itself as a lack of innovation in geothermal technologies. The problem is circular in nature: because geothermal energy is not seen as cost effective due to siting and technical difficulties, there have not been significant advancements in a number of years. But in order for costs to be considered reasonable, research and advancements must occur.

Fortunately, geothermal energy is quite mature. It has been around for a number of years and methods such as hot dry rock, steaming, and binary cycles work well. Of course, none of this matters to many potential investors due to the prohibitive costs involved in geothermal energy, even if a suitable location is found. The methods have been tried and tested multiple times, but without cost reductions, these methods cannot be improved, nor can new methods be discovered to possibly take the place of their predecessors.

Yet another hindrance is that no preliminary tests can be easily executed on a given location in order to test its potential for suitable geothermal energy stashes. In Chapter 5, we learned that hydropower devices are commonly so large that models span entire rooms or lakes. Similarly, a well must be dug in order to test for high yields of geothermal energy. Excavation, drilling, and surface temperatures must all align for a site to be labeled as acceptable; otherwise, all preliminary procedures will have been a waste of time and money.

Duration is yet another critical factor in determining the efficiency of a potential geothermal energy site. Once the well has been dug, drilling has begun, and necessary temperatures have been found, it is difficult to say how long a particular resource will last. A site could be built, tested, and deemed fully operational, only to have its geothermal energy run dry within weeks, months, or a few years. Geothermal is not a "renewable" energy source.

Also, recall that even a high temperature of 100 degrees Celsius reaches only a minimum level of what is required for a geothermal energy site to run efficiently. The drawn out process of a binary cycle is required for temperatures ranging from 100 to 150 degrees Celsius, causing many site inspectors to look for a minimum of 150 degrees Celsius, with higher temperatures being preferable. A temperature must fall within a certain range for the site to be worth exploiting.

7.6: HIGH POINTS OF GEOTHERMAL ENERGY

In the case of geothermal energy, the cons do outweigh the pros, but geothermal energy's positive points carry much merit. While geothermal energy is not exactly a "renewable" energy source, its means are not often easily depleted. Most geothermal energy sites will produce excellent results for years, and its efficiency ratings of 45 to 90 percent are sometimes all that is needed to convince a potential investor of a site's potential. Costs will hinder geothermal energy sites, but as long as operations run smoothly and efficiency ratings remain high, even an average site could negate crippling costs after a short period of time. Geothermal electricity is very dependable, and unlike solar power, it is not dependent on sunlight, nor is it as fussy a source as wind, nor as potentially damaging as hydropower.

Environmentalists should be pleased to know that geothermal energy is environmentally friendly, though not nearly to the degree of solar or wind power. Some disposal issues can become messy, but such instances are rare.

GETTING YOUR HANDS DIRTY -
CASE STUDIES

As fun as it is to learn new things, most nonfiction texts are stuffy. Quotes, dates, graphs, terminology -- everything is presented in such a stoic manner. I did not want this to be the case with Renewable Energy Made Easy, so now that you have trudged through the "stuffy" chapters, I thought you might be interested in having some fun.

This chapter introduces several projects from real people who have strived to commit themselves to renewable energy sources and their many benefits. In this chapter, you will learn about Cape Wind, an offshore wind power site. You will participate in hands-on projects, such as building your own solar panel, creating hydrogen, constructing a biogas generator, putting together a wind turbine, and more.

This chapter is easily my favorite part of the book. Anyone can read about a subject and claimed to have learned more about it, but the real test (and the real fun) comes from getting dirty with hands-on, do-it-yourself projects that better showcase how renewable energy sources are used in our everyday lives.

A special thanks to the individuals and companies who

contributed their time and projects with the intent of furthering your knowledge of renewable energy: Shane Jordan of **www.TheSietch.org**; Kathy Worobec of the Pembina Foundation for Environmental Research and Education (**www.GreenLearning.ca**; **www.Re-energy. ca**); Mark Rodgers and his colleagues at Cape Wind (**www.CapeWind.org**); Chris Graillat and the California Energy Commission (**www.EnergyQuest.ca.gov**); Michael Burghoffer; Simon Dale; and Doug Kalmer.

CASE STUDY: BUILDING YOUR OWN SOLAR THERMAL PANEL

Solar panels are perhaps the most ubiquitous form of understood, tangible, renewable energy. Everyone seems to have seen at least one, whether it is a glass-coated building (hopefully with functioning light wells) or a simple skylight that provides warmth in a family's den. When solar energy is discussed, solar panels are typically the focal point of the conversation simply because the concept is so easy to grasp: panels that store the energy beamed down from the sun that can be subsequently converted into a useable form of energy, such as electricity.

Yet despite their apparent commonplace status, not many people know how to use solar panels in their everyday lives. They are fantastic tools for heating water, buildings, and generating electricity, and surprisingly, many of the ingredients necessary to build your own solar thermal panel can be found in scrap heaps, hardware stores, and maybe even your own home.

Shane Jordan of **www.TheSietch.org** is a proponent of renewable energy and its copious benefits. The Sietch is founded on the principle of sharing information about renewable energy so that those without a background in mathematics and physics can successfully apply renewable energy in their everyday lives. With Shane's help, we will learn how to build a solar thermal panel in two different stages. The first stage details a complete solar thermal panel, but at a higher cost than stage two. Why is stage one included if stage two is more cost efficient, you ask? Shane was eager to provide both of his solutions.

For more information on The Sietch, be sure to visit **www.TheSietch.org**.

BUILDING YOUR OWN SOLAR THERMAL PANEL -- VERSION 1.0

Why Use Solar Thermal Panels?

"One of the main uses of energy in the American home is to heat up water," says Shane Jordan. "It takes a lot of electricity to heat up water; this costs a lot of money. All day while you are at work, or at school, the sun is busy heating up things outside. If you are a little kid and you get to play outside all day, then you know all about this. Lucky for us, the sun is free; all we need to do is figure out a way to use that energy to heat water. That's why we are going to build a solar thermal panel; these guys are great at heating up water."

Gathering Materials

The materials necessary for the construction of a solar thermal panel are many, but most can be found in junkyards, curbside trash heaps, and, if all else fails, almost any hardware store. If you are having trouble tracking down any of the following supplies, visit a store with knowledgeable staff, such as Lowe's or Home Depot, and they will be able to provide assistance.

For this project, you will need:

- Water

- Drill (make sure you have drill bits and screw bits)

- Scissors

- Any type of saw (hand saws will work fine)

- A 2 X 2 piece of plywood, one-quarter inch in thickness

- Two 2 X 1 sheets of Plexiglas

- Nine 2 X 2 base board planks measuring 20 inches in length

- One 5-gallon bucket

- One four-foot length of standard three-quarter-inch heater hose

BUILDING YOUR OWN SOLAR THERMAL PANEL -- VERSION 1.0

- One three-foot length of standard three-quarter-inch heater hose

- Four marine-grade stainless steel clamps

- Four, three-quarter-inch thru-hull PVC devices

- Two pieces of 2 X 2, black pond liner material (you can buy this in big roles or patch kits)

- Box of wood screws

- Wax paper

- Silicon caulk or that silicon goop stuff (like glue)

- Pond liner adhesive tape (this and the liner you can get at a water garden store)

If you read the introduction, you already know that this version of the solar thermal panel is a bit more expensive than the second version, but it is still only around $50 or so. This price is assuming that most people will go to hardware stores to purchase the necessary materials.

In terms of the time commitment you will need to make for this project, Shane estimates approximately three to five hours. "Once you get everything cut and assembled," Shane explains, "it only takes about [an] hour to set it all up."

Building the Solar Thermal Panel: Version 1.0

"Cut the backing and the pond liner," Shane instructs. When you are finished, you should have a 2-foot by 2-foot square of wood (one-quarter-inch thick) and two two-by-two foot pond liner squares. Drill a pair of three-quarter-inch holes in the wood as shown. Cut the outside slats as well.

BUILDING YOUR OWN SOLAR THERMAL PANEL -- VERSION 1.0

As you can see, the black pond lining sucks up all the light, making it look like some sort of black void in pictures. This is good because most of that sun goes towards heating your water. Next, cut three of your 20-inch baseboard planks with a little notch in the end as shown. It does not matter how big you make your notch, but try not to make it bigger than one-third the length of the board.

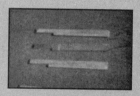

These boards will be used on the inside of the panel to direct the water in a wandering pattern. We want the water to go back and forth so that it will collect plenty of the sun's heat.

Now, the commencement of actual building. First, we have to pass our thru-hulls through the back of one of the layers of pond liner. To do this, cut a tiny "X" where you want the holes to be and press them through. Then coat the area around them with caulk or silicon gel and let it harden.

It may be hard to see in this picture, but the two thru-hulls are put in the upper-right and lower-left corner. They then fit into the wood backing (Into the holes you made before. Better make sure you line them up well or your backing will bunch up).

Next, you have to take the pond liner tape and cut it in half the long way.

You can no longer see the tops of the thru-hulls because the top layer has been firmly taped on. Next, we need to make the outside edge of the box, do this by screwing

BUILDING YOUR OWN SOLAR THERMAL PANEL -- VERSION 1.0

four of the 20 inch planks together. Then place it around the lining. It should fit snugly, but it is fine if the lining goes up the side a bit. When done, screw the little washers onto the thru-hulls.

Next, we take the slats we cut notches out of and screw them onto the sides. Place them over the tapelines you made in the middle, but be sure when screwing them in that you do not puncture the lining. You may want to put a bead of silicon goop around the edge and let it set up before going any farther.

As you can see, the slats match up with the lines of tape you made. Next, take the protective coating (it is usually blue) off of the Plexiglas sheets and then pre-drill holes into it so that it will not shatter when you screw it down; also pre-drill holes into the wood base, making sure not to puncture the pond liner, then screw the sheets on. Do not screw down too much or you will shatter the plastic.

It is almost impossible to see the Plexiglas in the picture, but it is there. Now we have to connect the hoses to the bucket and the back of the panel. Lastly, take two 20-inch

BUILDING YOUR OWN SOLAR THERMAL PANEL -- VERSION 1.0

wood pieces and add "legs" with a single screw at the top of your panel (one thru-hull should be up and one down).

Drill two three-quarter-inch holes into the 5-gallon bucket. Then pass the thru-hulls through them with the bung end facing out.

Put the washers on tight, as this keeps the water in the bucket. A bit of silicon cannot hurt either. Next, attach the longer hose to the bottom bung and the shorter hose to the top one; use the marine clamps to keep them on nice and tight.

Do the same with the panel (short hose on the top, long on the bottom) and clamp them down well. Now it is time to go outside and test it. Take your panel to a nice sunny place and set it up so that it is on level ground. Now fill the bucket up with water enough to cover the bottom hole about two inches above the hole.

BUILDING YOUR OWN SOLAR THERMAL PANEL -- VERSION 1.0

Face your panel into the sun with the legs and wait.

The water from the bucket flows down the long hose, into the panel, and then heats up from the sun. Hot stuff goes up, and water is no different. It is called "thermo siphoning." The water in the panel heats up and moves up the little path you have made for it; in the process, it gets very warm, so be careful. Then it pours into the bucket and starts its journey all over, in this way, you could make a large quantity of hot water if you put the hot water outlet into something else and kept putting cold water into the bucket. If you leave it in the sun, it will start to thermo siphon fast as the water gets increasingly warmer. It takes about four hours of sun to get going good, and on a sunny day, you can heat up a bucket to over 100 degrees easily, so be careful you do not burn yourself.

Enjoy the hot water.

BUILDING YOUR OWN SOLAR THERMAL PANEL -- VERSION 2.0

What Went Wrong?

Admittedly, the first solar thermal panel that Shane built works just fine. We are building another to learn. Just like researchers who work on refining renewable energy technologies each day, the best way to gain a deeper understanding of renewable energy and how it works is to learn what improvements were made and what they improved.

In addition, Shane felt that there were simply cheaper and more efficient ways to go about building a solar thermal panel. "After our first attempt at a home-built proof-of-concept solar thermal panel, we were a bit disappointed with the results," says Shane. "It took about four hours before the thing started work, and was a bit costly (at over 50 dollars) to make.

"I knew it could be done better and cheaper. My first mistake with the first one was purchasing everything new. With ample reusable resources at the local town dump, I knew it could be done on the cheap.

"Another flaw from the first panel was using pond liner as our collection medium. Pond liner is plastic, does not absorb heat as well as other materials (like metal), and is harder to work with, as you have to use glue or tape to create an air pocket to hold the water. It leaked the first couple of times we used it and took extensive repairs to make it work. We solved the problem by using a ready-made collector - something

BUILDING YOUR OWN SOLAR THERMAL PANEL -- VERSION 2.0

that was already designed to distribute heat, and made of metal.

"The last major flaw in our first panel was using Plexiglas for the cover. It is hard to work with as it will crack, and using two pieces left a hard-to-close crack in the middle. We solved this problem by using good old fashion window glass."

Gathering Materials

- Water

- 2 buckets

- Drill (with both drill bits and screw bits)

- Scissors

- A saw (a simple hand saw will do)

- Some wood

- A pane of glass

- The back of a small refrigerator

- 12 feet of air pump hose used in fish tanks

- Backing material (we used an old door mat)

- A box of wood screws

- Aluminum Foil

- Roll of duct tape

- Angle Cutter (or hack saw)

Instead of an estimated three to five hours total project time, Shane says that this version of his solar thermal panel weighs in at around three hours flat, and it cost less than $5 for him to make. What follows is Shane's own account, with pictures generously donated, of how he put together arguably the cheapest solar thermal panel ever made.

BUILDING YOUR OWN SOLAR THERMAL PANEL -- VERSION 2.0

Building the Solar Thermal Panel -- Version 2.0

Our local dump has a coolant removal program that has refrigerators and dehumidifiers that they remove old Freon from. With this in mind, I found the perfect heat collector. The back of a fridge is a heat dispersal system, and with a slight modification, it can be used to collect large amounts of heat.

Make sure that the Freon, or other coolant, has been removed, and cut the grill off at the base, near the large coolant holder.

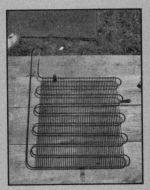

There was an old couch that had been run over by one of the large dump plows, and the inside wood was the perfect size for the frame. I found a pane of glass and an old rubber doormat that made the perfect backing and front.

BUILDING YOUR OWN SOLAR THERMAL PANEL -- VERSION 2.0

The glass was a real find, and it may be the only part of the panel that you need to purchase. Make sure your glass is big enough to fit over your collector and have enough room to attach it to the frame. The doormat was huge, so I had to cut it in half. There was a lot of nasty black goo, and a metal sheet in the middle. Remove the metal plate (or cut it in half as well) and leave the goo.

I built the frame around the collector, leaving enough backing to hold it all together. The frame is held on by building a similar frame on the back and driving large wood screws through the front frame, the backing, and into the back frame.

I added some foil to the backing. The reason for this is that, counter to what you would think, you do not want the backing to warm up. You only want the collector to absorb heat (it was so nice of the fridge company to paint it black for us). The foil will take any sun that was not absorbed by the collector on the first pass and bounce it back over the collector for another try at absorption. The glass cover will keep all the heat inside the panel for further absorption.

Light can pass through glass, but heat cannot.

BUILDING YOUR OWN SOLAR THERMAL PANEL -- VERSION 2.0

Notice how duct tape was used on the inside to seal all cracks; you could use caulk, but I did not have any, so I used the cheapest option. It worked well, and held the foil in place.

Next, we cut some notches for the entry and return ports to the collector.

Note again the use of duct tape to seal cracks.

I got some air pump hoses from the local fish store and attached them to the end of the entry and return ports. The duct tape was applied to make sure it was a tight fit; it was later removed, as it was not needed.

BUILDING YOUR OWN SOLAR THERMAL PANEL -- VERSION 2.0

Next, we attached the collector to the backing, using the mounting brackets that came on the fridge and some duct tape. If you wanted, you could use some screws and wood, but I found the tape and the natural tension of the construction to be enough to hold it in place.

Lastly, we attach the glass to the top. This serves to trap all the infrared radiation from the sun inside our panel where our collector will absorb it. Again, light can pass through glass, but heat cannot.

As you can see, simple duct tape is enough to hold it on. I would recommend using some sort of mounting bracket, however, as after a couple days in the sun, the tape started to droop, allowing the glass to slide off. A few screws would solve this, but I am cheap so I just put new tape on. Set your panel up at an angle so that it catches the most sun.

Here is the gross part: Put one end of the hose into your bucket of cold water, and make sure it is at the bottom of the bucket; next, grab the return hose and start sucking. Unfortunately, you have to prime the panel by getting some water into it. This can be done without getting water in your mouth, but inevitably, I sucked just a little too hard and ended up with a mouth full of nasty water. I would recommend

BUILDING YOUR OWN SOLAR THERMAL PANEL -- VERSION 2.0

having a friend do this part.

Set your cold-water bucket (source) up higher than your warm water bucket (return) and the whole thing will gravity siphon. Due to the design of this collector (both ports return to the same location on the panel), it will not thermo siphon. For that to happen, I would need to cut the long return pipe and have it exit at the top of the panel.

A word of warning, this panel works very well. We tested it on a very sunny day, and within seconds, the water coming out of the panel was hot enough to scald. I burned my fingers. This very hot water is only formed when the water inside the panel is allowed to sit for about a minute without moving. If the water is moving (due to the gravity siphon), the water exiting the return pipe is about 110 degrees, and, while hot, it will not burn you.

The water does not flow through the panel very fast (as the pipes are very small), but that a good thing as it allows the water to heat up significantly on its journey through the collector. It does take a while to heat up a 5-gallon bucket of water; I ended up building an insulated return bucket that was all black and sealed on the top, except for the port where the water tube enters. This kept the returned hot water hot long enough to be of use.

I let this system run for a couple of hours one hot, sunny day and heated up a 5-gallon bucket of cold water (measured at 70 degrees F) to over 110 degrees F. The temperature that day was about 76 degrees F. If the water was allowed to sit in the panel for several minutes and then forced out (by blowing in one of the hoses) the water was measured at 170 degrees F. Overall, we are much happier with the performance (and cost) of this panel. It performs much better than the previous one.

Our next modifications to this design will be to alter the return port so that it will thermo siphon, in this way, the return hose can be fed into the source bucket and the water will continually circulate in the panel, getting increasingly hotter. We have also talked about adding mirrors to the panel to concentrate more heat. Our goal is to boil water. This entire project cost less than five dollars, as I already had the screws and the duct tape. The only thing I purchased was the air hose, which cost $3.76.

CASE STUDY: BUILD YOUR OWN SOLAR BATTERY CHARGER

Shane Jordan of **www.TheSietch.org** spoke with me at length about how to create a solar-powered battery charger. Once it is complete, you could simply leave your batteries on the charger and wait for them to be recharged. No electricity would be used, and the environment would benefit from two less batteries rotting in a landfill.

Shane Jordan, of **www.TheSietch.org**, is a proponent of renewable energy and its copious benefits. The Sietch is founded on the principle of sharing information about renewable energy so that those without a background in mathematics and physics can successfully apply renewable energy in their everyday lives. For more information on The Sietch, be sure to visit **www. TheSietch.org**.

A word of caution for younger readers: Make sure to have an adult assist you with this project.

Why Build a Solar Battery Charger?

"Why pay for the batteries and the charger and then for the electricity to charge them?" Shane said in response to my question. "Let's build a solar battery charger instead. Then we can charge our batteries over and over and never have to spend a cent for that electricity."

Gathering Materials

- Old window (got mine out of a dumpster)

- Solar cells

- 22-26 gauge wire

- Duct tape

- Soldering iron and solder (use the stuff for fine electronics, not the heavier stuff)

- A cardboard box

- A AA battery holder (cheap at radio shack)

CASE STUDY: BUILD YOUR OWN SOLAR BATTERY CHARGER

- A blocking Diode (also from radio shack)

- A multi-meter (can be gotten cheap at radio shack)

The whole project was done on a Sunday morning, and it only took about two and a half hours.

Building a Solar Battery Charger

First, we need an old window; I got mine from a construction site. It is your basic old basement window. Make sure you wash both sides of the glass well with soap, water, and window cleaner, as more dust means less light gets through to your solar cells.

Next, take your multi-meter and sit in a sunny spot (or under a flood lamp) and check all your cells to make sure they get about .5 volts per cell. You do this by gently (cells are very fragile) putting the red on the front (make sure it is on the little metal lines) and black on the back (any place) of each cell while it is in the sun. It should say about .5 volts. If you get a bad one, do not use it, as it will lower the power of your charger.

For this project I used 12 cells, but you only need 6. So if you are on a budget use 6, but if you want a fast charger, use 12. Next lay out all your cells on your window to make sure everything is going to fit. I used the "deeper" side of the window so that there would be more room for a backing and because the other side has a shallower well that would cast less of a shadow on my cells if the sun moved while I had my panel lying out.

As you can see, they all fit with plenty of room left over; if I wanted to I could have used this window for a larger project, but I only need a battery charger for right now.

CASE STUDY: BUILD YOUR OWN SOLAR BATTERY CHARGER

Next we get down to the fun part: soldering. For anyone who has not done this before, it is easy once you get a little practice. Just make sure you open the window and let some air in because the room can get a little fume-filled. So first, a little bit about solar cells.

Solar cells are always .5 volts; no matter how large or small they are, they always pump out half a volt. The size only determines how many amps you get out of a cell.

There are two ways to wire power supplies (like batteries or solar cells) together. You can wire them in series or in parallel. If you wire them in series, you "add" the volts. If you wire them in parallel, you "add" the amps.

To wire in series, you wire the positive wire from one power source into the negative wire or the other. To wire in parallel, you wire the positive to the positive and the negative to the negative.

This might seem confusing, but it is easy. Most rechargeable batteries are 1.2 to 1.4 volts. To charge the battery, you need to pump more into it than is coming out. I want to charge two batteries at a time (1.2 X 2 = 2.4 volts), so a 3 volt array is what we need.

On a solar cell, the positive side is the "front" (the blue, black, shiny side), and the negative side is the "back" (the not blue, black, shiny side).

So let us make one 3-volt array. To do this, we need six cells. (.5 X 6 = 3). Solar cells are very easy to break, so handle with care. First, pick three cells and wire them like this, back to front, back to font. As you can see, the bottom cell has two pieces of wire soldered from its back to the next ones front, and so does the second one.

Here you see a close up of the cell with two wires attached to the front. I am using a special kind of "flat" wire, but you can use the round kind, as I will show you later.

CASE STUDY: BUILD YOUR OWN SOLAR BATTERY CHARGER

It is pretty easy to solder things to a solar cell; they will not melt. Put your wire down, touch your hot iron to it, and then touch the solder to it and pull your iron off. You will get the hang of it after the first couple.

Make sure you solder to the wire lines that are already on your cell. These lines are put there to allow the electrons from the panel to flow into your load. If you do not get a good attachment to them, your charger will not work. After you solder three of them together going "back to front," you have to do a second set. This time, you need to go from "front to back," because these three will connect to your first three. This is the "back" side of the cells.

As you can see, the second set of three (at the top) are done "front to back." If you look at this picture, you can see that if we laid all six of these cells out in a row, it would be very easy to connect all of them in series, negative to positive, repeatedly. But our window is not big enough to hold all of them in a long line, and it would be harder to get sun to shine on something that long without shadow, so we are going to put them into a stacked pattern.

Right now, what we have are two 1.5 volt arrays, but we want one 3-volt array. So we need to connect the "front" or one array to the "back" of the other. We do this with some wire.

When you are done, you basically have a 3-volt array, which will charge your batteries just fine. But if you want more amps (faster charge), you need another 3-volt array. So get another six cells and do all this over again. When you are done, you should have two 3-volt arrays, like this:

CASE STUDY: BUILD YOUR OWN SOLAR BATTERY CHARGER

Notice the four leads I have soldered on to the back, I made a mistake here (top points if you can spot it). What I did wrong was I put the leads correct for the front, but I failed to remember that like a battery, you cannot put both leads on one side. You need to put the positive on one side and the negative on the other.

After a bit of solder removal and redoing, it is all ready to go. Therefore, what you have now are two 3-volt arrays that have a positive (front) and negative (back) wire coming off them. To wire them in parallel (and thus increase the amps), you simply twist the two positive wires together and the two negative wires together. The two positive wires become one + wire, and the two negative wires become one - wire. Attach the + wire to the red lead of your radio shack AA battery holder and the - wire to the black lead from your battery holder and you are almost done.

Affix the blocking diode to the negative lead that goes to the battery. This will keep current from flowing back out of the battery when a cloud goes over head or the cells are shaded.

Right now, our cells work just fine (if we connected them all correctly); put them in the sun and you should get 3 volts with good amps, but if we try to pick up our cells, they will all fall off and break. So we need to attach them to the glass somehow. I went low-tech on this one: good old duct tape.

CASE STUDY: BUILD YOUR OWN SOLAR BATTERY CHARGER

All I did was duct tape each cell and then put some parts from a cardboard box on the back. Feel free to make your backing more weatherproof (water + solar cells = bad). This charger is going to be put in the back of my car or on a windowsill, it is not meant to be put out in the weather. If you want yours tougher, make it so.

If the sun is too bright, you are going to want to watch your charger; too much charging and your batteries can be ruined. On bright days, it is a good idea to let your batteries charge for only an hour, then take them out and see whether they are full. You can do this with a multi-meter.

Have fun with your freely charged, rechargeable batteries.

CASE STUDY: PREPARING DELICIOUS FOOD AND DRINK WITH SOLAR ENERGY

Leaving conventional gas-powered or even electric ovens unattended while away is dangerous, though some people do make the unfortunate mistake of leaving them on, sometimes with disastrous results.

Fortunately, Shane Jordan of **www.TheSietch.org** has a solution for all of you plagued with poor memories -- a solar-powered oven. With this simple device, you really can afford to leave the oven unattended while you are out running errands, working, or socializing, and you can even use it as a legitimate excuse if you find yourself stuck in an awkward or dull situation. Of course, given that the solar oven will be powered by the sun, you will have to find a different excuse to leave your family reunion should it take place after sundown.

As delectable as your solar-cooked food is sure to be, you will need a drink to go with it. Shane has a solution for this as well: solar tea. With solar-cooked food and solar-warmed drink, you can leave the house for several hours and enjoy a complete, delicious meal upon returning, without worrying about burning your house down while you are away.

Part I: Building a Solar Oven

Your conventional gas or electric oven surely prepares your food more than adequately. In addition, those ovens take less time to prepare food than your solar oven will. Saving energy is the goal of using a solar oven. Shane Jordan feels that

CASE STUDY: PREPARING DELICIOUS FOOD AND DRINK WITH SOLAR ENERGY

his (or any other) solar oven is not meant as a complete replacement for your conventional kitchen appliance, but rather, a substitute to use when you are not concerned about taking a little extra time to prepare your food. A solar oven is perfect, he says, if you want to place your food in the oven and then do something else to occupy your time for a few hours. Have errands to run? Go do them. Want to watch a movie? Watch away. When you are finished and ready for food, your solar oven will have done the job nicely, and without using electricity or gas.

"The idea of a solar oven is that you use the sun's power to cook your food," Shane says. "You basically make an insulated box and cover it with glass. Then we use mirrors to bounce more sun into our box, [which] makes it even hotter."

Gathering Materials

- An old window. Shane recommends keeping an eye out for this in dumpsters, junkyards, or junk put out by a curb.

- Some plywood. Again, turn to dumpsters and other trash heaps to avoid spending money.

- Duct tape.

- Black spray paint.

- Cardboard boxes.

- Some nails and a hammer. Alternatively, screws and a screwdriver can be used if those are closer at hand.

- A hand saw.

- Some old mirrors. Foil and cardboard works well as an alternative if you happen to have these products around the house.

The total time investment for this project: approximately 4 hours.

Building Your Solar Oven

What follows is a direct account from Shane Jordan on the exact process he used to construct his solar-powered oven.

CASE STUDY: PREPARING DELICIOUS FOOD AND DRINK WITH SOLAR ENERGY

First, we need an old window. I got mine from a construction site. It is your basic old basement window. Cut off the outside wood part so that you can get the glass inside. Take your saw and cut slowly on one side until you hit glass, then remove the wood.

Be wary of the sharp glass.

Now comes the "hard" part. By hard, I mean it will be the most physical effort. You need to take your wood and make a box. What size wood you have will determine what size box you make. You want to make your box big enough to hold at least one pot. If you make a box that is too small, it will not fit your pot, and if you make it too big, it will not get hot enough because there will be too much air to heat up.

I originally made a huge coffin-like structure (we at The Sietch get a little excited about these things). I found that it just would not get above 100 F, no matter how big my reflectors. I then cut it in half to a box 14 inches high, 26 inches wide, and 17 inches deep.

A bit of measuring, cutting, and nailing later, I had a nice little box made. I then went and raided the local dumpsters for cardboard.

One thing you can count on is that there are always going to be many cardboard boxes lying around. Take your duct tape and line all the inside seams with tape. This will help to keep heat from getting out between the boards. Take your cardboard and line the inside of the box until you have about 1 inch of cardboard on the sides and bottom of the box.

I found that it was easiest to use large pieces of cardboard and jam them in tightly instead of trying to use many smaller pieces and tape or glue them in. Pay special attention to the top of the box; you want to make sure the cardboard is all one level so that it makes a good seal when we lay the glass on top of it later.

When you are done, you should have a nice looking box with a smooth top. As you

CASE STUDY: PREPARING DELICIOUS FOOD AND DRINK WITH SOLAR ENERGY

can see, I have painted the box black (the tape was later painted black also). Make sure that you use paint that says "lead-free and nontoxic when dry" on the back.

I had a great find the other day while going through dumpsters: a mirror. I find these all the time, but they are usually broken. If you are not lucky enough to find mirrors, you can use foil and cardboard (use a mix of white glue and a little water to affix it to the cardboard). You can paint the inside black as well to collect even more heat.

You might want to clean your mirror better than I did. Now here is the part I am going to leave to you, you need to find some way to affix the mirrors to your box so that you can reflect the sun into the box. I used some wood and rebar that I found to rig up a holder that sits separate from the box, but the sky is the limit.

Some tips for solar cooking -- put your food out in the morning with your reflectors facing south (or if you live in the southern hemisphere, point it north), and when you get home, it will be ready to eat. Most things take about twice the recipe cooking time. But the nice thing is you cannot really overcook in a solar oven, so once its cooked, your food will just stay hot and ready to eat. Also, if you can paint your pot black and put it inside an oven bag, this will make it get hotter faster and stay hotter longer.

You only need a day that is sunny 20 minutes out of every hour. If you can find some sheet metal, paint it black and put it in the bottom of your box; it will collect heat and help to keep your food hot .It can go from ambient temp to 120 in about 10 minutes, with cooking temps in the 250-275 range.

A word of warning: your solar oven might not produce enough heat to thoroughly cook meat. Use a regular oven for such items.

Have fun cooking with the sun.

CASE STUDY: PREPARING DELICIOUS FOOD AND DRINK WITH SOLAR ENERGY

Part 2: Making Sun Tea

"So it is summer, and it is hot out," says Shane Jordan. "You want to have some nice tea. The last thing you want to do is heat up the house by turning on the oven. Lucky for you, there is a giant ball of thermonuclear fire 10 light minutes away that has plenty of heat to spare."

Gathering Materials

- Water

- Tea

- Plastic Wrap

- Sugar

- A rubber band

- A container

- Sun

The total amount of time you can expect to invest in this project is a convenient one hour.

Making Delicious Sun Tea

What follows is Shane's account of how to make sun tea.

Making sun tea is pretty easy. First, find yourself some tea, some sugar (if you like sweet tea), and a container. Read the side of the tea box. Our tea said about 1 cup of water for each tea bag, so measure out enough water for how much tea you want to make, then tie the tea bags together and drop them in.

You can then put the top on, using a bit of cling wrap and a rubber band. This lets plenty of sun in the top and holds the tea bags up. Then, find a sunny spot and set your tea out. I used this electric do-dad because the top gets hot and is not in the shade.

CASE STUDY: PREPARING DELICIOUS FOOD AND DRINK WITH SOLAR ENERGY

The tea starts to "tea up" right away. At this point, it is best to go play in the grass, see a movie, or do something fun. When you come back, your tea will be done.

Note the dark rich tea color. Take it inside and prepare to taste; for me that means plenty of sugar and some lemon. Some like it chilled with ice; however you like it, you can know you saved a little bit of carbon dioxide by using the sun instead of the stove.

CASE STUDY: CREATING YOUR OWN HYDROGEN

Shane Jordan of **www.TheSietch.org** has not only invested his valuable time in the world of renewable energy, but he is also interested in general science and how it connects to renewable energy. When studying hydropower, Shane thought it would be interesting to share his method of creating custom hydrogen molecules.

A word of caution for younger readers: Make sure to have an adult assist you with this project.

Why Create Your Own Hydrogen?

You may have heard of the upcoming "hydrogen economy" and wondered what all the fuss is about. What everyone is talking about is just how clean hydrogen is as a fuel. Hydrogen produces only water vapor and oxygen when it is burned (used) in a fuel cell. The only problem is that because hydrogen is the lightest of all elements, it will only hang out in nature attached to other elements.

CASE STUDY: CREATING YOUR OWN HYDROGEN

The first thing we need to know about water is what it is made of. Water is H_2O, meaning it has two hydrogen atoms and one oxygen atom. Water is mostly hydrogen. This is good for us because we are going to be busting it up with a little electricity.

What we are going to do is called "electrolyzing" the water. When we send electricity though water, it gives the atoms of hydrogen and oxygen enough energy that they break free from each other and shoot out of the water (the hydrogen and oxygen gas is less dense than water, so it shoots to the top).

Gathering Materials

- Water

- A power source (we will be using solar panels, but you can use a rechargeable 9-volt battery)

- An old lantern battery (non-alkaline)

- Wire

- Salt

- Sand paper

- Wire cutters

It took me a couple of weeks to track down an old battery (you could just go buy one, but why waste the cash when they are free at the dump), the solar panel took me about two hours to build, but the project itself only took about 20 minutes to complete.

Creating Hydrogen

The first step is to figure out what kind of power source you are going to use. In the most technical terms, you could use power out of the outlet for this. Just get a cheap extension cord, cut the end off, and shove both wires into water. This, however, is a very bad idea. There is possibly nothing more foolish than to stick something plugged into the wall into water. Therefore, to make this project much safer, we are going to

CASE STUDY: CREATING YOUR OWN HYDROGEN

use one of two power supplies.

The first option is to use a rechargeable 9-volt battery. Nine volts is more than enough to break up water, and because the battery is rechargeable, we can keep using it repeatedly. However, it can take several hours to recharge a battery, and you will only get about 15 minutes of useable power from one. For this reason, I am going to use solar panels. I built a 12-volt solar array.

For instructions on how to create your own solar panel, please see "Case Study: Building Your Own Solar Thermal Panel" on page 150.

Hook up two wires to your array, one on the positive lead and one on the negative lead. When you electrolyze water, you just stick two wires into the water and the bubbles form. The more surface area you get touching the water, the more bubbles will form. That is why we are going to use carbon rods instead of plain wires. Most people do not have carbon rods lying around the house (you can use pencil lead, but it is hard to get out of the wood, and it is not as pure as the ones we are going to use). To get carbon rods, we need an old lantern battery.

You may have one of these sitting around the house right now, begging you to use it for something other than land fill. First, open the top. Not to worry: old carbon-zinc batteries like this do not have acid inside that will splash out and hurt you.

CASE STUDY: CREATING YOUR OWN HYDROGEN

Leave the little wires on the top, as they will come in handy later for connecting your carbon rod to your power source. Next, we need to open up the canisters. Now if you are concerned with chemical stuff, this is the time to put some rubber gloves on. But when you cut the top of the canisters off (I just used some wire cutters), you will see hard, black, clay-like goop, wrapped in some kind of paper, and inside that you will find your carbon rod. Here is what it looks like at each stage, with the rod on the left.

Do not be alarmed when you start removing the black goop, it will be messy and flake all over the place. Be sure to have a bag or box to put all the insides in. I did this whole part of the project outside to avoid mess. Your hands will be covered with black stuff when you are done, so take a hand-wash break.

You can remove the plastic "necklace" that is still on your rod (leave the little copper "hat"). Then scrape all the goop off it with a knife or X-acto blade. Then take a bit of sandpaper and rub the rod down to make it nice and smooth. When you are done, you will see that you have what is in essence four large pencil leads (made of carbon, not lead).

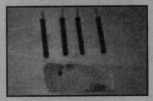

The rods are very interesting; you can draw with them like a pencil, but let us save them for busting up that water. Strip the ends of the wires on your rods so you can

CASE STUDY: CREATING YOUR OWN HYDROGEN

attach them to your power source. If you are using the battery, attach the wires right to the battery (be careful that you do not attach a wire across the top of your battery, or let the positive and negative wires touch each other, as it can heat up and burn), then attach them to the rods. If you are using the solar array, attach the wires from the leads to the rods.

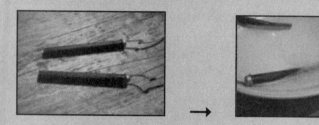

We have moved outside on a nice sunny day, and our rods are all attached. Lay your solar array some place it will get plenty of sun and you are ready to go. Get a bowl of water and add some salt to it. The salt makes the water more conductive, making more of our electrons bust up more water molecules. Water is mostly hydrogen, so you will see that one of your rods makes many more bubbles than the other. The one that makes more is hydrogen. You can test it by holding a lit match over the water with the hydrogen bubbles. They will burn; this is safe, so do not worry about making an h-bomb or anything. They will just flare up a bit.

This rod is making oxygen...

...while this rod is making hydrogen.

Notice that more hydrogen comes out (H_2O, hence more hydrogen in water). You can even collect the hydrogen if you want. I used a tube with a balloon on the end.

CASE STUDY: CREATING YOUR OWN HYDROGEN

If you do collect the hydrogen, there are a couple of safety tips you need to know. If you put the rods close together and collect bubbles from each, you will be collecting the perfect explosive gas mix. Oxygen and hydrogen gas, when mixed together, is very flammable, so be careful. Also, if you put much salt in the water, there is the possibility you might create a bit of chlorine gas, this is poison and should not be breathed in; be sure to do your experiment in a well-ventilated area. If you are using a solar array, this will be outside, so you are OK.

If you do this experiment in bubble soap, you can create hydrogen bubbles. I was unable to do this, as I did not have any bubble soap lying around. You are able to light the bubbles on fire and make a very loud noise. This was done for me in science class once, and it shook the room. It was loud but harmless, as hydrogen burns very fast and cleanly (leaving only water vapor and oxygen).

The hydrogen created by this method is pure enough to use for many fun experiments; it is possible to let this process run all day long and collect a large amount of hydrogen. Have fun making your own hydrogen -- I did. It was fun to see how much gas comes out of just plain water. It is also fun to put your hands in front of the solar array and watch the output of the gas go down. Try it for yourself.

CASE STUDY: BUILDING YOUR OWN BIOGAS GENERATOR

The concept of biofuel is an intriguing one. Each of us disposes of organic waste every single day, so why not learn how to put it to good use? It is an easy suggestion to make in theory, but putting it into practice by way of finding a biofuel plant can be quite difficult, depending on your location from the plant. As we discussed in Chapter 3, biofuel plants located far from a proper fuel source is one of the main detractors of biofuel as a renewable energy.

If a biofuel plant is too far away to be practical, why not bring the plant to you? The Pembina Foundation for Environmental Research and Education has graciously provided detailed instruction that enables readers to build their very own biogas generator. Just like building a solar panel out of common materials, constructing a household biogas generator is simple and inexpensive; all you need to donate is a bit of time.

Re-energy.ca is a part of the GreenLearning family of education resources of the Pembina Foundation for Environmental Research and Education (**www.greenlearning. ca**; **www.re-energy.ca**). GreenLearning produces sustainable energy and climate

CASE STUDY: BUILDING YOUR OWN BIOGAS GENERATOR

change education resources and is managed and delivered by the staff of the Pembina Institute.

Finding and Gathering Required Materials

To build your own biogas generator, you will need the following tools and materials:

Tools

- Tubing cutter

- Scissors

- Adjustable wrench

- Rubber gloves

- Electric drill with a quarter-inch bit, or a cork borer

- Hot glue gun with glue sticks

- Electrical or duct tape

- Sandpaper, or a metal file

Materials

- Used 18L clear plastic water bottle

- Large Mylar-based helium balloon

- A plastic water bottle cap with "no-spill" protection

- Copper tubing that measures 40 centimeters long and 6.5 millimeters inside the diameter of the tube

- T-connector for plastic tubing; barbed, 6 millimeters, or a quarter-inch long

- One tapered cork measuring 23 millimeters long

- Clear vinyl tubing measuring 1.5 meters long and 4 millimeters or a quarter-inch inside the diameter

CASE STUDY: BUILDING YOUR OWN BIOGAS GENERATOR

- Two barb fittings, each a quarter-inch by a quarter-inch

- Quarter-inch ball valve

- Six to eight manure pellets; retrieve pellets from goat, sheep, llama, rabbit, or another similar animal

- Thick pair of rubber gloves

- Large plastic tunnel; this can easily be made from a four-liter plastic milk jug with its bottom removed

- Wooden stick measuring 30 to 50 centimeters in length, and 2 to 3 centimeters in thickness

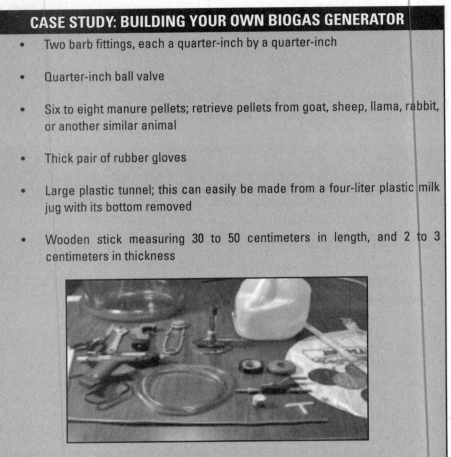

Once you have gathered your materials, you are ready to begin building your very own biogas generator.

The easiest material to locate should be the water bottle, which is commonly carried in most grocery and hardware stores that sell purified water. These stores typically collect bottles that can no longer be refilled due to being damaged or considered too dirty for reuse. The best part: the bottles are usually free of charge due to their unusable states. To inquire about collecting as many bottles as possible, go to you nearest grocery store and ask to speak to a supervising manager.

Though not commonly known as "Mylar balloons," these balloons are quite prolific in everyday life. Remember those pitch-black balloons your friends got you when you turned "old?" Yes, those very same balloons are made of Mylar, a type of strong yet thin polyester film used in functions such as photography and insulation. To track down proper Mylar balloons, visit any party supplies store; some florists carry them as well.

CASE STUDY: BUILDING YOUR OWN BIOGAS GENERATOR

For materials such as t-connectors and copper tubing, barb fittings, corks, and similar products, a visit to versatile hardware stores such as Lowe's or Home Depot will be in order.

Warning: this part of the material gathering process stinks -- literally. If you know someone who keeps domesticated animals such as rabbits, cows, llamas, or other animals that produce pellets, gathering the goods will be a (smelly) breeze. For those who do not have friends who work professionally with dung, many types of manure can be bought at gardening supply stores.

Taking Proper Safety Precautions

The main hazards in this activity are from sharp tools, such as tubing cutters and scissors. Exercise caution while using any tool. There is no risk of explosion due to the leakage of methane because the gas develops so slowly that it dissipates long before it can reach flammable concentrations in room air. Exercise the normal precautions in the use of Bunsen burners: keep hair and clothing away from the burner while it is lit.

How Will Our Custom Biogas Generator Work?

The apparatus you are going to build uses a discarded 18-liter water container as the "digester." A mixture of water and animal manure will generate the methane, which you will collect in a plastic balloon. The 18-liter water container performs the same task as the stomach of a livestock animal by providing the warm, wet conditions favored by the bacteria that make the methane.

Building Your Very Own Biogas Generator

The first step in the process of building our biogas generator is securing the biogas collection system.

1. Cut a 20-centimeter piece of copper tubing. Round off the sharp edges of the freshly cut tubing using sandpaper or a metal file.

2. The Mylar balloon has a sleeve-like valve that prevents helium from escaping once it is filled. This sleeve will help form a leak-proof seal around the rigid tubing. Push the tubing into the neck of the balloon, past the end of the sleeve, leaving about 2 centimeter protruding from the neck of the balloon, as shown below.

CASE STUDY: BUILDING YOUR OWN BIOGAS GENERATOR

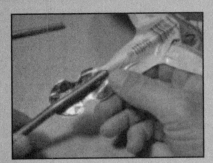

Carefully insert the copper tubing into the neck of the balloon.

3. Test the tube to be sure air can enter and leave the balloon freely, by blowing a little in through the tube. The balloon should inflate with little or no resistance, and the air should be able to escape easily through the tube.

4. Securely tape the neck of the balloon to the tube as shown in the illustration.

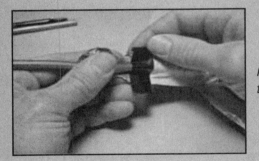

Make sure that the neck is securely taped.

5. Using a drill or cork borer, make a small (4mm) hole in the center of the stopper. Add a few drops of hot glue around and inside the hole and insert the stem of the quarter-inch T-adapter into the cork.

Carefully apply glue to the cork.

CASE STUDY: BUILDING YOUR OWN BIOGAS GENERATOR

6. Screw the two barb fittings into the body of the ball valve. Tighten with the adjustable wrench.

Use the adjustable wrench to install the barb fittings.

7. Cut two sections of vinyl tubing, each 25cm long. Use them to connect the balloon to the T-adapter, and to connect the ball valve to the Bunsen burner. Assemble the rest of the gas collection system according to the diagram below.

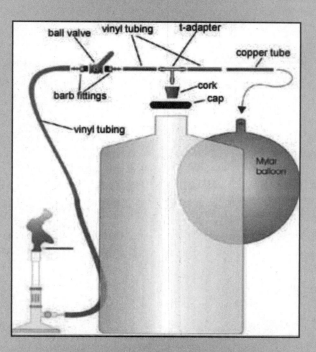

Now that the biogas collection system has been carefully built and constructed, it is time to move on to Step 2: preparing the manure mixture. The Pembina Institute wisely recommends that this part of the process be carried out outside and with

CASE STUDY: BUILDING YOUR OWN BIOGAS GENERATOR

durable rubber gloves.

1. Cut the bottom off a 4-liter plastic milk jug to make a wide-mouthed funnel.

2. Place the funnel into the neck of the plastic water bottle and scoop in small amounts of manure.

Grab a friend and play your favorite process-of-elimination game to determine which one of you will have to get down and dirty with the manure. Hint: Play "Eenie, meeny, miny, moe" and start off pointing at yourself.

3. Here is where things could potentially become even messier. Should the neck of the bottle become plugged, use a stick or piece of dowelling to push the manure through.

4. Add enough water to bring the level close to the top of the water bottle.

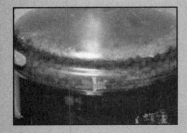

Watery manure. This is what renewable energy is all about.

5. Use the stick to stir up the manure and water mixture, releasing any bubbles of air that might be trapped.

6. After working with such messy, hazardous wastes, it is important to remove

CASE STUDY: BUILDING YOUR OWN BIOGAS GENERATOR

your gloves and clean up carefully. Use soap and disinfectant to wash your hands as thoroughly as possible.

Now that the most disgusting piece of the puzzle has been fitted into place, the remainder of our biogas generator's construction should proceed smoothly -- and cleanly. Take all of your materials and arrange them on a clean, clear, flat surface, and continue to follow the instructions.

1. Snap the cap onto the top of the manure-filled 18-liter water bottle.

2. Be sure the ball valve is closed, but also make sure that the gas moving from the water bottle can pass freely through the T-adapter to the balloon. After this step is complete, your completed biogas generator should look similar to the following:

Congratulations. You are now the proud owner of a genuine biogas generator.

3. Place the biogas generator in a warm location. Locations such as over a heat register or radiator or in a sunlit window are considered the most ideal. A word of caution: If the biogas generator is placed in a window, be sure to wrap the outside of the container in black plastic or construction paper. Protecting the biogas generator in this way will help to discourage algae from growing inside the bottle.

Now that you are finished, it is time to test your biogas generator. Follow these steps carefully and safely. Younger readers should be sure to have an adult present in order to handle matches or a lighter.

1. First, open the clamp or valve so that biogas can flow back from the balloon to the Bunsen burner.

2. Have a friend squeeze the Mylar balloon gently while you attempt to light the Bunsen burner with a match or spark igniter.

3. If your Bunsen burner ignites, your biogas generator is a success.

CASE STUDY: BUILDING YOUR OWN BIOGAS GENERATOR

Testing Your Biogas Generator

Like all forms of technology, your biogas generator might initially work just fine, but eventually simmer out. However, do not be alarmed if it appears that the generator is not working as planned the first several times you use it. For the first few weeks, the biogas generator will produce carbon dioxide. As discussed in Chapter 3, this is a negative drawback to biofuel, but do not fret: once all of the oxygen inside the bottle has been used up by aerobic bacteria, the anaerobic bacteria responsible for making methane will take over.

Remember, it was explicitly stated at the outset of this Case Study that time and patience are two very important investments. Your custom-built biogas generator could take as long as a month to make biogas with enough methane to be flammable. The result will be a functioning renewable energy generator and the knowledge that you have done something positive in the fight to preserve fossil fuels by using a powerful renewable energy source.

CASE STUDY: DOUG KALMER'S SOLAR-POWERED HOME

It is common for many modern locations to use various forms of renewable energy. Scattered throughout almost any given community, you will find buildings with glazed windows for space heating, photovoltaic arrays, wind farms, and cars able to rely on biodiesel. Many people apply renewable energy in scattered forms, but what about a structure completely powered by renewable energy? How would such a building perform all of the functions that any other building might use?

Quite simply, if you ask Doug Kalmer. In 1985, Doug built a solar-powered house in Tennessee. This ambitious project came from successful encounters he had had with varying forms of renewable energy in the past. "I bought a 2,000 square-foot, un-insulated house in upstate New York, which got me interested in solar space heating," Doug explained. "After the usual insulating and weather-stripping, I added an attached greenhouse and built a window box solar space heater. These worked well enough to make me want to explore further solar possibilities. After reading everything I could on the subject, I decided to build a passive solar earth sheltered home in Tennessee. I researched the subject by reading magazines and books and visiting every solar building I could."

In 1982, Doug's project began to take tangible form. Join Doug as he takes us on a solar-powered journey that ends at a house equipped with a greenhouse, a custom

CASE STUDY: DOUG KALMER'S SOLAR-POWERED HOME

masonry stove, and a solar water heater.

My wife and I bought 34 acres in Tennessee, and started to build our dream home --
passive solar space and water heating, earth sheltered, post and beam framed, slip-
formed stone, with cedar cordwood in-filled south wall. We had some money from
selling a house I remodeled, but funds were tight, so we did all the work we could
ourselves. We hired a track loader to excavate a 25' by 65' recess into a south facing
hillside, and then hand dug footer trenches, poured 15 yards of concrete in them, and
started slipforming stone walls. We placed locally gathered stone into the forms and
mixed concrete with a gas mixer to fill around the stone. Once set up, the form could
be moved, using the same forms over and over again to move down the walls.

I then felled and ripped posts and beams from oak trees on our property. I also cut
Eastern Red Cedar (Juniper) into 16" lengths to stack for drying. After framing and
decking the original shed roof, we poured another 15 yards of concrete with the
help of friends and neighbors for a slab. We floated the slab surface for texture, and
stained it a dark brown to improve solar absorption.

After 15 months of drying time, we could wait no longer for the cedar to dry further,
so we started building with it. First I built a shaving horse and using a drawknife,
I had to peel all of the bark from the cedar. I later learned that if I had cut the
cedar in the spring, when the sap is up, it would of peeled more easily. Laying two
strips of sawdust rich mortar along the inside and outside of the wall, I laid the 16"
cedar lengths onto them, filling the inner space with insulation. This way there is
no continuous mortar bond through the wall to transfer heat. The cedar cordwood
wall was labor intensive, but cost little cash to build. Everyone likes the way it looks
and smells. However, soon after moving in we noticed drafts where the wood meets
mortar- air infiltration. When the wind blew rain through the wall, I knew I had to
cover the outside of the cedar.

CASE STUDY: DOUG KALMER'S SOLAR-POWERED HOME

I first used 6 mil plastic, covered with hand split oak shakes. It helped with the infiltration, but insects found the shakes provide a good home, several types of wasps, even a bat moved in. I finally removed the shakes and plastic, tacked 5/8" closed cell Styrofoam board over the cedar, and stuccoed over the entire outside wall. This stopped the insects and infiltration. I now do not recommend cordwood walls for dwellings, as the rate of expansion/contraction with humidity changes is very different for mortar and wood, infiltration is inevitable, unless you tightly cover the outside wall.

Since 1985, I've been living in a solar collector-otherwise known as a direct gain passive solar home. It is naturally well lit, thanks to many large, evenly spaced windows on the south wall. These appropriately shaded windows allow direct sunlight to reach the back of the building in winter, but allow no direct sunlight inside in summer. The light which does enter strikes the textured, brown concrete floor, slip formed stone walls, and large stone fireplace, gently warming these surfaces which absorb and store heat, moderating temperature fluctuations.

Having insulation on the exterior of the building allows these thermal masses to remain at or near room temperature, absorbing heat during sunny days and radiating warmth at night. This makes interior temperatures very stable, naturally staying warm in the winter and cool in the summer. Because the floor and walls are doing double duty as thermal flywheels, temperatures also remain very even throughout the house.

His simple system is effective enough to require backup heat only after cloudy days in December, January, and February. My only backup heat is a large stone fireplace, modeled after the high thermal mass Russian and European designs. Mine also provides domestic hot water. My space and water heating bills are near zero. Passive solar systems are simple in concept and use, have few or no moving parts,

CASE STUDY: DOUG KALMER'S SOLAR-POWERED HOME

require little or no maintenance, and require no external energy input. Passive systems collect and transport heat by non-mechanical means.

Active systems employ hardware and mechanical systems to collect and store heat, often using some outside energy source such as electricity for fans and pumps. The greenhouse effect is most commonly demonstrated by leaving a parked car in the sun with the windows rolled up. We all know how hot that can get because it lacks storage. In a passive solar building, your windows are your collectors, your walls and floor are your absorbers and storage. Water and phase change materials can also provide storage. Typically one half to two thirds of total surface area is masonry. An open design aids heat and light distribution.

A direct gain building is the simplest live-in solar collector-heat storage and distribution all in one. They work well on sunny days and in cloudy climates by collecting and using every bit of energy that passes through the glazing, direct or diffuse. Masonry thermal storage materials include concrete, concrete block, brick, stone and adobe.

Indirect gain is when sunlight first strikes a thermal mass, located between the sun and the living space, commonly called a Trombe wall.

An attached greenhouse is a combination of direct and indirect gain.

Isolated gain is when collector and storage are isolated from the living space. Temperature fluctuation can be controlled by operable windows or vents, shading devices, and a backup heating system. Windows can still receive 90% of possible gain when oriented within 25 degrees east or west of true south.

Incorporating solar design into a new structure is fairly simple and low cost. My passive solar, earth sheltered, post and beam framed house cost about $8 per square foot to build, not including labor. There are no special materials required- just a rearranging from typical design. Looking at ordinary homes, I usually see picture windows facing the road, when it would have been a simple change initially to orient to the south, providing a lasting energy gain in winter, instead of a loss. In the northern hemisphere, south facing windows provide a net energy gain; all other directions are an energy loss.

Also, many homes are built with brick or stone on the outside of the insulated shell, when a reversal of positions, with insulation exterior to the thermal mass, would greatly improve the homes' thermal stability and comfort at little or no added cost. Having heat absorbing masonry materials as part of the home's interior structure,

CASE STUDY: DOUG KALMER'S SOLAR-POWERED HOME

such as floors and walls, can reduce overall building costs, especially when, if things are designed correctly, the need for a furnace or boiler can be eliminated, or at least downsized.

Sometimes there are minor problems with having sunlight entering your home. At times I find a certain chair too brightly lit for comfort, but I just move to another. This is the advantage of spreading the windows out along the southern wall. You have some solid wall in between windows, to minimize glare and provide some shaded areas. I suppose the sunlight also helps fabrics fade, although I haven't noticed this occurring. People in more populated areas may have some privacy concerns with a lot of large windows facing their neighbors, but this can be designed around, possibly going to a Trombe wall, or indirect gain system.

Even though it's simpler to incorporate into a new structure, a lot can be done with existing structures. Energy conservation must be the first step, since there is little use collecting solar heat if you can't hold onto it. Weather-strip and insulate, possibly adding exterior insulation. Consider moveable insulation for doors and windows at night. A small addition of a double door or airlock entry way can increase energy efficiency, and give you a place for all those shoes and coats.

An attached solar greenhouse, or sunspace, can provide heat, food, beauty, and additional room. Plants thrive in them. My 8' x 18' attached solar greenhouse cost $250 to build, and my wife enjoyed it and what it can do for plants so much that we now have a 22' x 48' freestanding greenhouse for her plant business. Properly placed vegetation is also important, even for houses with no solar aspect. Deciduous trees, shrubs or vines on the east, west, or south sides will lose their leaves in winter to allow sunlight in, while providing cooling shade in the summer. Evergreen foliage on the north side will buffer winter winds.

Solar hot water can be added to existing structures, as I did to my house 17 years

CASE STUDY: DOUG KALMER'S SOLAR-POWERED HOME

ago. I am now well past the point where the money I invested in the solar water heater equals the money I would have spent on electricity to heat water. Consider the fact that in the next five to eight years you are going to pay the cost of a solar water heater, whether you buy one or not. It's your choice-you can invest in solar now, demonstrating your support for sustainable energy, and getting free hot water after your payback period, or continue to pay ever-increasing energy bills, which indicates your support for maintaining the status quo.

Passive solar design is not just about heating. Many solar design considerations also help with summer cooling. Thermal mass resists overheating, direct earth contact through slab-on-grade, and earth sheltering all contribute to cooling in hot weather. The most efficient shape of building for maximum winter solar gain is elongated along the east-west axis, giving a large south facing wall and smaller east and west facing walls. This design also minimizes unwanted summer heat gain on the hot east and west sides.

Radiant barrier placed in the attic or roof system can reflect 97% of radiant heat, keeping the excess solar gain in summer from the living spaces. Light colored roofing also helps. Vegetation is usually the best shade, because it is later arriving in the spring, when we need more solar gain, and usually provides shade into fall, as well as proving its own evaporative cooling effect.

There are several low-cost, low-tech devices that anyone can use, such as an integral passive solar water heater which is basically a tank in a box. Window box collectors, window greenhouses, and attached greenhouses can help heat the house. Solar food dryers, cookers and ovens can also reduce utility bills. Many of these can be homemade, inexpensively.

I woke up to 4 degree Fahrenheit outside, and two of three indoor thermometers read 65 degrees Fahrenheit, and the third read 68 degrees Fahrenheit. With a short but sunny winter day yesterday, high of 33 degrees Fahrenheit, warmed it up to 76 degrees Fahrenheit inside, it felt good. This morning, another 4 degrees Fahrenheit reading, and 65 degrees Fahrenheit inside on two out of three readings, interior most thermometer reads 68 degrees Fahrenheit.

This is the most extreme conditions we face-short winter days, cold nights, and my direct gain, low cost (<$10 sq per square foot) passive solar home stays comfortable enough, and I now know that I could of done better 25 years ago when I built it. I used 2" of expanded foam board as my exterior insulation, and went to only 1" at 4' or more below ground. I didn't protect the bead board from moisture or termites, so

CASE STUDY: DOUG KALMER'S SOLAR-POWERED HOME

I'm sure it's degraded by now, yielding less than it's original R3 per inch. I know I have infiltration at some windows and doors. I have some large windows that don't have nighttime insulation. Yet, the masonry floor and walls store enough solar heat from the short day to get us comfortably thru the long cold nights, even with these heat leaks.

Yes, the house went through a 11 degree Fahrenheit temperature swing, but that's very livable for us, I like it cooler at night for sleeping, and that's the most extreme temperature swing it goes thru. More commonly, winter temperature swings are under 10 degrees Fahrenheit. This is accomplished by having lots of surface area of masonry thermal mass directly exposed to sunlight, which is much more effective for storing heat than mass not in direct sun. We have not burned any wood during this extreme cold (for us) period, and wood is our only backup heat.

Some typical questions that I get about my solar home are:

- **Doesn't it get hot inside in the summer?** No, proper orientation and shading prevent sunlight from entering the building during the summer months, keeping it cooler than the average home.

- **Doesn't it cost more to build?** No, properly done it can cost less.

- **Does it have to face south?** Yes, in this hemisphere it does-within 25 degrees east or west of south.

We speak of "producing" oil, as if it were made in a factory, but we don't produce oil; all we do is mine it and burn it up. Neglecting the interests of future generations who are not here to bid on this oil, we have been squandering, in the last few decades, an inheritance of hundreds of millions of years. Only recently have we begun to come full circle to the same realization that similar boom and bust cultures have reached before us: that we must turn back to the sun, and seek elegant ways to live within the renewable energy income that it bestows on us.

As sure as the sun will rise tomorrow, our energy costs will also continue to rise. Getting heat from sunlight is economical, ecological, dependable, readily available, time tested, powerful and empowering. This free and equally distributed energy source arrives at our homes almost daily. Let's all try to make better use of it, for our own well-being as well as the planet's.

CASE STUDY: DOUG KALMER'S SOLAR-POWERED HOME

Doug's Masonry Stove

Living [through] 10 cold New York winters solely on wood heat made me want to learn more about wood burning stoves. I studied and tried out several different designs. Airtight stoves appeared to be the most efficient, but they are very smoky and inefficient when damped down to hold a long, low fire for hours without tending.

Because my house design works well, wood heat is now my only backup heat source, and we burn only a small amount of wood yearly. But during those cold times, extra heat is needed. I designed and built my own version of a masonry heater (also called Russian masonry stoves, Finnish masonry stoves), which was developed in cold northern climates. These stoves are typically built with 20-30 tons of stone mass, thus holding their heat for days.

Here in central Tennessee, where it's rare to have extended cold, cloudy periods, I want heat quickly, but not four days after I fire it. By designing an air space between the firebox and stonework , along with room air vent openings high and low in the

CASE STUDY: DOUG KALMER'S SOLAR-POWERED HOME

stone surround, I get much faster warm-up than the traditional masonry stove. Still I get the moderating and heat saving features from the mass of the stone surround.

My stove also heats domestic water for my 85-gallon solar water tank in the attic, which adds to the thermal mass for solar heat storage.

To improve the draft of the stove I insulated the exterior of the masonry chimney. If you online search "masonry stove," you find many expensive kits and plans explaining the basic theory of masonry stoves. I used local, free stone that I gathered for the 3-foot deep by 9-foot wide by 9-foot tall surround, which averages about 10-inches thick. That's about 10-12 tons of stone. It's on a 6-inch thick slab, poured separately from the house slab.

I welded a firebox from scrap 1/4" and 3/8" steel, making an airtight door with a tempered glass window in it. I ran 4" galvanized vent pipe from the attic down through the surround, to the back of the firebox, and made a channel in the concrete slab for the incoming air to flow under the fire to the front glass, washing across the glass to keep it clean. Using air from the outside also prevents the air needed for combustion from pulling replacement cold air in around our home's weather-stripping on doors and windows.

I used 4 feet of 6" stainless steel exhaust stack from the firebox, which then goes 2' over into a block-and-tile chimney. I coiled 30' of 3/4" copper tubing around the vertical section of stack to heat water; it flows by gravity to my solar tank located in the attic directly above it. The tank is supported by the stone fireplace.

There is no damper, nor a thermostat to regulate the fire, so it burns hot and cleanly.

CASE STUDY: DOUG KALMER'S SOLAR-POWERED HOME

In 20+ years of use, I've not needed to clean the chimney yet. The heat is soaked up into the stone and concrete. It takes about an hour for a pulse of heat to travel an inch through stone, so we get a nice warming in back of the fireplace after the fire itself is cooled to ashes.

With the solar house, we burn little wood, but I'm still glad I went for the masonry stove. It is efficient, esthetically pleasing, and ends the typical problem of either being too warm or too cold, as with a regular stove.

Doug's Solar Greenhouse

My attached greenhouse is 8'x18', and attached to the south side of my solar heated home. I built both in the mid 80`s, for little money. The greenhouse initially cost me about $300 to build, but it was single glazed then.

Started out by pouring a footer, and laying up 6" blocks, which I then externally covered with 2" of foam board, topping it off with an 8" treated sill plate. I had found some 46"x76" sliding glass door replacement tempered glass for $15 each, so I framed walls with 2x4`s to fit them.

I initially used corrugated fiberglass for the roof, it lasted about 10 years, then I replaced it with twinwall polycarbonate, which is much better. I have a pea gravel floor over the soil, a scrounged brick pathway, and four 55 gallons drums of water for thermal mass. The drums support benches, and are along the house wall.

I put in a planting bed and shelving for plants, and my wife found out she liked growing plants so much it started her out in a greenhouse business. The attached greenhouse serves to start plants in January, then in February we transfer them to

CASE STUDY: DOUG KALMER'S SOLAR-POWERED HOME

the 22'x48' freestanding greenhouse.

The attached also is an airlock in the winter, as it covers an entrance door, which we open in sunny cold weather to help heat the house. It is a handy place for firewood, and the dogs stay there on winter nights. I also installed an exhaust fan thru the adjoining wall , which I wired to an AC thermostat, so it can come on at about 85 degrees Fahrenheit greenhouse temperature to blow warm air into the house, useful when we are not home.

The adjoining wall also has a window with a lower vent space below it, when we are not going to be home during sunny cold weather, I removed the insulated cover to the vent, so air can be returned to the greenhouse from the house, it has a 4 mil flap on the greenhouse side to prevent reverse flow. I have large screened vents in both east and west ends for warm weather ventilation. I made extruded foam board inserts to block them off in winter. In summer I cover the roof with 60% shade cloth.

The greenhouse provides heat, acts as an airlock, provides a place to grow food and house plants, it helps us earn a living by providing a place to start plants to sell, it is a great place to hang out in sunny cold weather, all in all, we really like it.

Doug's Do-It-Yourself Solar Water Heater

This Photovoltaic (PV) pumped hot water system has been working well, with no maintenance, for years on my house. I am now well past the point where the money I invested in the solar water heater equals the money I would have spent on electricity

CASE STUDY: DOUG KALMER'S SOLAR-POWERED HOME

to heat water. Most of the year, we have more free hot water than we can use. Consider the fact that in the next five to eight years you are going to pay the cost of a solar water heater, whether you buy one or not. I kept costs down by doing all of the work myself, and buying a used collector panel, but still created a long lasting, efficient, high quality system.

I started by looking in the yellow pages under "Solar," and found a plumber in the nearest large city, with spare collector panels. I bought a 3`x13` aluminum collector with copper tubing and tempered glass cover for $100, guaranteed not to leak, but I tested it first, anyway. Assuming the collector needs to be freeze protected, a non-toxic antifreeze, usually propylene glycol, is pumped through the collector, to the heat exchanger tank.

Collector mounted on the roof. PV panel to drive circulator pump is visible at the upper left.

Solar water heater plumbing running through the attic. The 12 volt DC circulator is mounted to the 2X4. The pump used is no longer available, but pumps offered by Ivan Technologies (El Sid) or March Pumps would be good substitutes.

CASE STUDY: DOUG KALMER'S SOLAR-POWERED HOME

The simplest way to circulate the anti-freeze is with a 12 volt pump directly wired to a small photovoltaic panel; this eliminates the need for controls of any kind. When the sun is heating water, the pump runs. I bought a matched set of a 10 watt 12 volt solar electric panel, and bronze magnetic drive circulation pump from Zomeworks.

A PV pumped glycol system is absolutely freeze proof, uses no grid electricity, infinitely adjusts flow thru the collector in perfect synchronicity to the heat in the collector, has no electronic controls to malfunction, only needs one small pump, is zero carbon, and is easier for a homeowner to install, as there is no concern about improper drainage freezing and damaging the system. The COP (Coefficient of Performance) is infinite on a PV pumped system, well above any system that requires grid energy. Full disclaimer- I am just a homeowner who has studied the options and lived with my system for 17+ years, I do not sell or install systems for a living, so I have no vested interest in any particular product.

I wanted my system to be totally freeze proof, off grid, and as simple as possible, so I chose PV pumped closed loop, which requires one small pump, which can be PV powered, while drain back needs two larger ones, which are grid powered. Differential controllers, and sensors are not part of the PV pumped closed loop system, these parts can be problematic. In a PV pumped system the sun is your control. There is a simple elegance to having the PV panel directly wired to the circulating pump, no further controls needed.

The sunlight reaching the collector panel varies exactly as the sunlight reaching the PV panel, so, if they are properly sized, as the sun sets, or clouds move in, the pump slows at the same rate the heat slows to the collector. The collector does not hold heat well; it's designed to transfer the heat to the small amount of fluid in the lines, which heats up quickly.

Just moving the water around a filled loop does not require much power or much energy, (because the water falling down one sideof the loop lifts the water up the other) but pumping the water up an empty loop is a sizable load on the systems energy usage, so either the number of times that happens, or the energy demand of each time, or both, should be reduced, if possible.

Closed loop antifreeze systems provide the most reliable protection from freezing. These systems circulate an antifreeze solution through the collectors and a heat exchanger. Propylene glycol is the most common antifreeze solution. Unlike ethylene glycol (used in automobile radiators), propylene glycol is not toxic. Closed loop systems like this are quite common, whether they be for solar domestic hot water,

CASE STUDY: DOUG KALMER'S SOLAR-POWERED HOME

radiant floor heating, or hydronic baseboard heating. Despite the additional parts and fittings, they have a high degree of reliability, and are well understood by heating contractors.

My years experience living with a PV pumped closed loop system tells me that it works well just as it is, remember, sensors, controllers, microprocessors, and their failures were a large reason for the demise of many SDHW systems in the 70's and 80's. Let's keep it simple, and reliable. Let the sun be your control.

He most expensive single component was the Heat Exchanger (HE) tank. I looked at what is available, and most are built around the ordinary glass lined steel water heaters, which have a limited lifetime. For less money, I had a stainless steel tank built with 50' of 3/4" soft copper tubing in the lower half as my Heat Exchanger. It holds 85 gallons, and acts as a pre-heater for the regular electric water heater. Non ferrous stainless will outlast me.

Proper sizing of the system is important. Plan on at least 20 gallons of storage tank size for each of the first four people and 15 gallons for each additional person per day. You should have at least 40 square feet of collector area for the first two family members, then add 12-14 square feet for each additional family member. Keep tank size at a ratio of 1.5 gallons or more to one square foot of collector area to prevent overheating.

Storage tank with internal heat exchanger

Besides the pump, panel, HE tank, there are a few other parts to the system:

1) A pressure gauge (0-60psi) will let you know the closed loop has not lost its charge of antifreeze.

2) A solar expansion tank allows the solar solution to expand as it heats.

3) A check valve above the tank to prevent thermosyphoning at night.

CASE STUDY: DOUG KALMER'S SOLAR-POWERED HOME

4) A pressure relief valve.

5) A hose valve at the lowest point for filling and draining. It really is a reliable, efficient system; we do no maintenance to it. I can read water temperature going in to my electric water heater, and most of the year, it is above the 120 degrees setting of the thermostat.

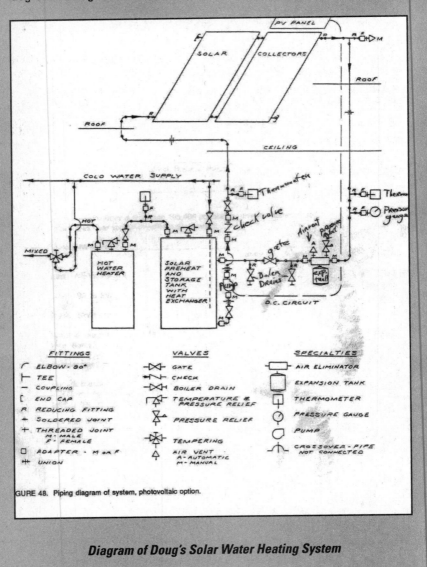

GURE 48. Piping diagram of system, photovoltaic option.

Diagram of Doug's Solar Water Heating System

CASE STUDY: DOUG KALMER'S SOLAR-POWERED HOME

Doug's Simple Thermosyphon Solar Water Heater

Doug's simple thermosyphon solar water heater. Collector to the lower left, and storage tank with wrap around heat exchanger on shelf in shop. No pumps!

It's been over 17 years since I installed a PV pumped solar hot water system on my house, and that success made me want to design a system for my shop/studio outbuilding. Because people saw the panel on the roof of my home word got out that I knew something about solar, I was called to remove a system from a neighbors house. It was an example of the systems from the early eighty's -- overly complex, used grid electricity to run two pumps, and this system was aluminum. I was paid to remove two 4x8' aluminum panels, and the pumps, heat exchanger (HE) and water tank. One panel has a leak, but the other checked out fine.

While the system on my house works well, I did have to invest in a 10 watt solar electric panel to run a small circulating pump, and have a stainless steel heat exchanger tank built. The total for the house solar system was $1100. The pump was necessary because the HE tank is in the attic below the roof mounted panel. Hot liquid will rise on it's own, but must be pumped downward. However in the shop, I could have the panel down low, and the Heat Exchanger tank up high, so I hoped to not need the expensive pump and PV panel.

Fortunately for my purposes, the aluminum hot water panel had a serpentine pattern, so the liquid enters on one end down low, goes the length of the panel, rises a few inches, and comes back, repeating the pattern. This works well for a thermosyphon system, as the heated liquid naturally rises, pulling cooler liquid in below. Technically, the cooler denser liquid is heavier, so it is pulled down by gravity, displacing the

CASE STUDY: DOUG KALMER'S SOLAR-POWERED HOME

lighter, hotter liquid, but for simplification, heat rises.

Now I just needed to figure out how to build a low cost Heat Exchanger. I bought a new 38 gallon electric water heater, the shorter style. I removed the metal ends and shell, promptly voiding the warranty, I'm sure. I used a wood chisel to remove two strips of foam insulation, below and above the thermostat. I bought a 50' coil of 5/8" copper tubing, and wound it around the tank as tightly as I could, using strap clamps to pull it into direct contact with the metal tank walls.

The idea is to keep the coils as low on the tank and in close contact with the exposed steel. I wrapped the coils in copper flashing and twisted copper wire around the flashing to hold the coils in as tightly as I could. Leaving the two tubing ends sticking out, I insulated the tank with radiant barrier, which is reinforced aluminum foil, then a layer of Reflectix, which is double bubble pack covered with Mylar. Both are available at building supplies stores; I just had some around.

Wrap around heat exchanger

Tank installed on shelf with insulation

I reinforced a heavy duty shelf in my shop and used the front end loader to put the tank on it. The bottom of the tank is about 6' above the top of the collector panel. I then used 5/8" automotive heater hose to connect the top of the panel (Hot out)

CASE STUDY: DOUG KALMER'S SOLAR-POWERED HOME

with the top of the HE coil around the tank, and the bottom of the coil to the bottom of the panel. This way, whenever the sun shines, the non toxic antifreeze (propylene glycol), heats up, rises to the top of the HE coil, gives up its heat to the water tank, cools, and descends back down to the bottom of the panel.

I then super insulated the tank with 2" thick foam panels built to surround it. I used about 20' of heater hose, and at first I thought I'd have to use an expansion tank to accommodate the expansion and contraction of the fluid with temperature swings, but I so far the heater hose allows enough movement to prevent too much pressure rise. It has been thru several long, hot summers, and has not had any problems with expansion or contraction of the working fluid. I have changed to cheaper, more readily available automotive type ethylene glycol, as the double walled HE eliminates any contamination concerns.

I put a tire valve at the highest point of the HE tubing, so I could bleed air from the system, and check pressure. I have about $60 in the solar end of the system, I would have spent the $160 for the electric heater anyway. The electric heater is wired up as back-up. The system works, but the weakness is the smaller size of the tubing (5/8") and the lack of a pump means the temperature difference has to be fairly great to circulate much heat. I am considering going back in to add cement around the HE coils, in direct contact with the tank, to improve the heat flow thru the HE and tank wall.

This information was re-printed with Doug Kalmer's permission.

CASE STUDY: SIMON DALE'S LOW IMPACT WOODLAND HOME

CLASSIFIED CASE STUDIES
directly from the experts

For fans of J.R.R. Tolkien's Middle-Earth, I imagine the above picture caused quite a bit of excitement. Though it may look like a hobbit hole, neatly tucked away in the Shire, the home is actually quite real and belongs to Simon Dale. Simon decided to

CASE STUDY: SIMON DALE'S LOW IMPACT WOODLAND HOME

demonstrate his belief in energy conservation by constructing a fully functional woodland home and graciously allowed for his carefully documented journey to be shared with *Renewable Energy Made Easy's* readers.

For more information regarding Simon and his "hobbit hole," visit his Web site at **www.SimonDale.net**.

Basic Overview

The house was built by myself and my father in law with help from passers by and visiting friends. Four months after starting, we were moved in and cozy. I estimate 1000-1500 man hours and almost $5000 at that point. Not really so much in house buying terms (roughly $95 per square meter excluding labor).

The house was built with maximum regard for the environment and by reciprocation gives us a unique opportunity to live close to nature. Being your own (have a go) architect is a lot of fun and allows you to create and enjoy something which is part of yourself and the land rather than, at worst, a mass produced box designed for maximum profit and convenience of the construction industry. Building from natural materials does away with producer's profits and the cocktail of carcinogenic poisons that fill most modern buildings.

Some key points of the home's design are:

- Dug into hillside for low visual impact and shelter

CASE STUDY: SIMON DALE'S LOW IMPACT WOODLAND HOME

- Stone and mud from diggings used for retaining walls, foundations, etc.

- Frame of oak thinnings (spare wood) from surrounding woodland

- Reciprocal roof rafters are structurally and aesthetically fantastic and very easy to do

- Straw bales in floor, walls and roof for super-insulation and easy building

- Plastic sheet and mud/turf roof for low impact and ease

- Lime plaster on walls is breathable and low energy to manufacture (compared to cement)

- Reclaimed (scrap) wood for floors and fittings

- Anything you could possibly want is in a rubbish pile somewhere (windows, burner, plumbing, wiring...)

- Wood burner for heating - renewable and locally plentiful

- Flue goes through big stone/plaster lump to retain and slowly release heat

- Fridge is cooled by air coming underground through foundations

- Skylight in roof lets in natural feeling light

- Solar panels for lighting, music and computing

- Water by gravity from nearby spring

- Compost toilet

- Roof water collects in pond for garden, etc.

Main tools used: chainsaw, hammer and 1 inch chisel, little else really. Oh and by the way I am not a builder or carpenter, my experience is only having a go at one similar house 2yrs before and a bit of mucking around in-between. This kind of building is accessible to anyone. My main relevant skills were being able bodied, having self belief and perseverance and a mate or two to give a lift now and again.

CASE STUDY: SIMON DALE'S LOW IMPACT WOODLAND HOME

CASE STUDY: SIMON DALE'S LOW IMPACT WOODLAND HOME

The Construction Process

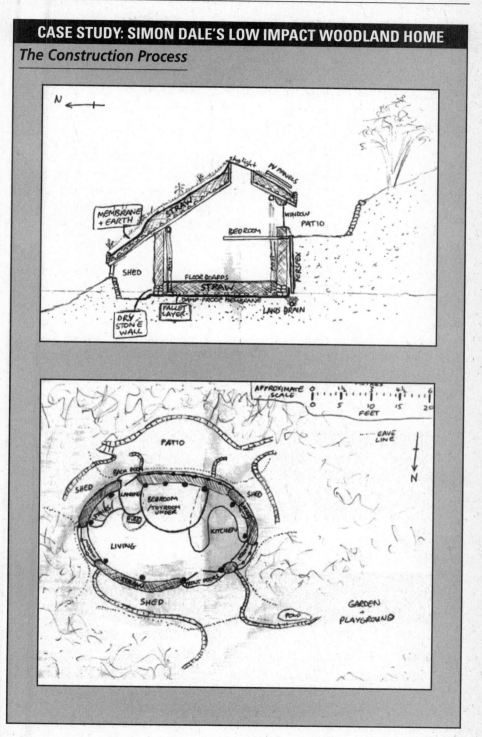

CASE STUDY: SIMON DALE'S LOW IMPACT WOODLAND HOME

Take one baby, a toddler, and a building site. Mix well with a generous helping of mud, combine with 6 weeks of solid welsh rain whilst living under canvas. Do this in candle light without a bathroom or electricity for three months. Chuck in living with your father for good measure. Top with an assortment of large slugs. The result a hand crafted home of beauty, warmth and health for about $5000.

Having children is a major motivation for buying a house. Combine the rigors of looking after young children and meeting demanding mortgage payments in today's climate and you have a recipe for stress. Then add to this concern about toxic materials that are inherent in most buildings, and exposing your precious babies to them is a cocktail of dismay. This would sum up the options before our family before we decided to take the plunge and go off the beaten track.

Some past experience, lots of reading and self-belief gave us the courage of our conviction that we wanted to build our own home in natural surroundings. With this is mind my husband and I decided to build ourselves what you might call an eco-home wherever we could get the opportunity. For us one choice led to another and each time we took the plunge events conspired to assist us in our mission. Looking back there were times of stress and exhaustion, but definitely no regrets and plenty of satisfaction.

Initially we had no capital and we had decided resolutely to be full time parents whilst our children were young. As you'll appreciate to be a full time mum and part time dad our income is low, so a mortgage was not an option and the prospects for renting seemed grim. Providence came our way and a landowner offered us the chance to move to his woodland in west Wales to build an eco-house. There would be no formal security or long term ownership, but about $3200 was available for materials, so we jumped at the idea without a backward glance.

So here we are today. You can see from the photos our home is unusual but the aesthetic appeals to lots of people and perhaps touches something innate in us that evolved in forests. We hope this article will provide confidence and information for anyone inspired to undertake a similar project or even just to illustrate that where there's a will there's a way even if such a building does not tantalize your taste buds.

Many people ask how we managed to build a house whilst camping without mainstream facilities and as the mum rather than the full time building blokes (my husband and father) I can assure you of a few things. Children like mud, diggers, tools, wood and candlelit extended camping. Mums fret about washing. Dads build

CASE STUDY: SIMON DALE'S LOW IMPACT WOODLAND HOME

all day long and then look after mum. Children are entertained by the outdoors ad infinitum even when it rains. Mums hanker after cozy cafes and make frequent excursions to venues with warm, clean toilets. Children find sticks, look under stones for insects, collect acorns, simulate diggers and do a lot of puddle splashing. Dads carry on building and look after family in the evenings when they are not completely exhausted. Children see materials taking form, observe the construction process and make a lot of connections; they see their parents being effective. Mums wash up whilst yearning for tiled utility rooms, learn to ignore the mud and fend off slugs. Dads build, console mum, read children bedtime stories and make muddy imprints on the sheets. Everyone wonders at the nature of slug slime. Then one day you get a house.

If that hasn't put you off more serious issues do provide an impetus to eco-building. Modern construction materials, mainly cement and insulation, contribute significantly to carbon emissions and pollution of water, soil and air albeit often in other countries. More alarmingly home furnishings and interior finishes account for the primary pollutants we are exposed to, much more so than from traffic fumes or factories. Fluoro-carbon compounds in fire retardants permeate nearly everything we live amongst; toxic chemicals such as bisphenol-A are rife in paints and varnishes; PVC windows, doors, kitchens and so on contaminate our air and devastate the environments they are produced in. All these substances are well documented as harmful to human and plant life, and cancer forming agents. Some are belatedly being phased out but what about their saturation of our world already? I have no answer to this except to choose natural materials when I can and am confident that cost does not have to be the prohibiting factor.

So what did the house that rose from mud, slugs, physical effort, three months and a few grand entail? A hole was dug into a bank with a knackered old digger (not my father, a machine the landowner already had on site) to the required depth for us to have a mezzanine floor. This took a few weeks interrupted by me constantly dragging men to the glorious West Wales beaches as it was mid August after all. A rough stone wall was built around the perimeter for the wall foundation. The stone was sourced from the hole and free.

The structure was oak posts used in the round, i.e. not sawn into planks and hence very strong. In our case thin, spindly oak posts were all around us in the woods and not conventionally useful for any purpose except shamefully firewood, fencing or wood-chip. This required a chainsaw, men, lots of heavy lifting, a few days and no money. To construct the frame some advice and calculation was necessary but this information is readily available from other people who have built 'roundhouses.'

CASE STUDY: SIMON DALE'S LOW IMPACT WOODLAND HOME

Please see link below.

DPC plastic was laid in the hole and salvaged hardwood crates laid on top to provide a breathing space for straw bale insulation. Eight tons of straw bales arrived on a huge lorry, costing about $1030, which were stacked in glorious sunshine and secured from the deluge of rain that began the next day and lasted for weeks. Very quickly the bales were assembled as walls, secured with hazel rods, leaving appropriate gaps for windows. Straw bale construction is extremely quick, easy, fun and hence rewarding. As our house is oval the bales require jumping on to achieve a curve.

The roof is a series of layers to achieve insulation, waterproofing and low visual impact with little cash. Cotton dust sheets were laid on to the rafters, followed by whole straw bales, sealed with three layers of silage plastic and topped with earth from the original digging. A space was left in the centre for a skylight.

Serendipity called again to provide the floor and windows. Large pine palettes from a nearby works due to be burnt were laid on the straw, sanded and polished up with a non-toxic oil and wax that we have found to be durable and great to work with. When you are oiling boards in the same room as your kids at 4am you are glad it is only citrus your hair and clothes are covered with rather than petrochemicals. Due to a north facing aspect we have no windows in the front of the house but designed the length of the rear wall to have windows. A nearby double glazing supplier gladly furnished us for free with misfit units due to be landfilled that miraculously fitted with an inch to spare and came framed with fittings.

By this time we were exhausted, I had evacuated the children to a luxury holiday cottage for a fortnight to escape the mud and now snow. The Dads were working round the clock to make the house habitable only stopping to eat sardines. They occasionally slept but generally carried on building the mezzanine, fitting back doors, digging grey water pits and plastering with semi frozen lime and fingers. All the tools were buried in straw. Dads were in danger of being mistaken as yetis. So in December 2005, sixteen weeks on, we moved in without doors but with a huge sense of satisfaction.

The time is now to make changes in our lifestyles and consumption. There are a number of projects making this a reality in Britain and Ireland and plenty of resources to support green building ideas, links to these can be found on our website, **www. simondale.net/house**. If this has inspired you to get covered in straw and lime, come and help us build another grander home next year! Feeling impotent in the face of environmental and social problems is overcome more easily than we imagine by

CASE STUDY: SIMON DALE'S LOW IMPACT WOODLAND HOME

forming clear intentions of our ideals. Realizing them is not always simple, but in our experience more fulfilling than business as usual. For peace of mind and the future we bequeath to our children let's do the job as best we can.

As dozens of green writers have already pointed out, we need new policies that conserve nature while encouraging people to choose and build their own homes. We need to reverse the flow of people from the land to the cities, and to give people something worthwhile to do. We must grow more of our own food, organically, and reduce dependence on fossil fuels and techno-fixes. Will there be humans living here in a thousand years? Birds, trees, hedgehogs, apples? If so, we have to move fast now.

CASE STUDY: BUILD YOUR OWN WIND TURBINE

When I was a child, I was fascinated in everything mechanical. It would be nice, I thought, to have a windmill in my front yard to provide electricity to my house. Unfortunately, they were a bit too big to fit in the bed of my stepfather's truck, so for awhile, the dream was dead.

Luckily, Michael Burghoffer of **www.PicoTurbine.com** must have been attuned to my inner desires. Michael's site features a kit filled with instructions and parts that enable consumers to build a smaller windmill-esque device called PicoTurbine. Michael's uncle initially built the PicoTurbine as a hobby before realizing the importance it would have to society. By building a PicoTurbine, Michael and his uncle hope that wind power will become more understandable and accessible to everyone.

Two options for acquiring the PicoTurbine exist. You can head to PicoTurbine.com and order your own kit, or you can simply gather the necessary materials listed in Section II and follow the step-by-step instructions detailed in Section III.

Ready to get to work? I thought so. Continue reading for the Burghoffers' step-by-step guide detailing the construction of your very own PicoTurbine.

About the PicoTurbine

This document will show you how to build the PicoTurbine -- a fully functioning, electricity-producing scale model of a Savonius wind turbine. The entire project costs

CASE STUDY: BUILD YOUR OWN WIND TURBINE

only a few dollars, using commonly available materials like magnets, cardboard, tape, wood screws, and a wooden dowel or pencil.

The PicoTurbine can be built in about one hour if you have the kit, about 90 minutes if you do not. The kit has the wire coils pre-wound which saves a lot of time. Less time is needed if done as a group project. With some adult supervision PicoTurbine can be assembled by children as young as ten years old, making it an excellent project for renewable energy education.

PicoTurbine stands about eight inches tall--but don't let its size fool you. This version of PicoTurbine produces about one third of a watt of power from a direct-drive single-phase brushless permanent magnet alternator. The design is naturally self-limiting for over-speed protection. I've left models out all night during a windstorm with 50 mile per hour gusts that made my brick house shake. In the morning I looked out my window--fully expecting to see it shredded--only to find PicoTurbine still spinning at top speed in the early morning gale!

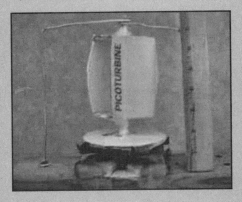

Gathering Materials and Reviewing Safety Rules

If you purchased a PicoTurbine kit from **www.picoturbine.com**, then you already have everything listed here. Otherwise, be sure to find:

- 2 feet of 10-gauge aluminum wire.

- 400 feet (about 1/3 pound) of 28 AWG enamel coated magnet wire.

- 4 ferrite magnets, about ¾ inch wide, 1.75 inches long, and 3/8 inch thick. Strong ceramic grade-5 magnets are recommended.

CASE STUDY: BUILD YOUR OWN WIND TURBINE

- 1 mini-lamp, 1.5 volt, 25 milliAmps.

- 1 bicolor light emitting diode (LED) with 2 leads.

- 3 Phillips head (cross slot) screws.

- A piece of wooden dowel ¼ inch diameter and 7 inches long. A pencil or long pen will work.

- An scrap of wood, 8 inches long, about 4 inches wide, and ¾ or more inches thick. A piece of 1x4 or 2x4 works well.

- A piece of corrugated cardboard about a foot square, perhaps cut from a box.

- Scotch tape and any type of glue.

You also need the following tools:

- Scissors

- Ruler

- Phillips head screw driver

- Pliers

- Pencil sharpener

It is also helpful to have the following tools, but not entirely necessary:

- A digital multimeter that can measure AC millivolts is useful for tuning and testing the alternator, and displaying the amount of electricity produced.

- Sandpaper is helpful for stripping enamel-coated wires, but if you don't have any handy you can carefully use the blade of your scissors against the side of the 2x4 wooden block.

CASE STUDY: BUILD YOUR OWN WIND TURBINE

PicoTurbine is not a dangerous project to build, but as with any construction project certain safety rules must be followed. Most of these rules are just plain common sense. Be sure to review these rules with children if you are building this project as part of an educational curriculum.

- Adult supervision is required for this project.

- This project is not recommended for children under 10 years old.

- Children must be supervised when working with scissors and other sharp parts to avoid cutting injuries.

- Children under 4 years old should never work with wire or small parts like screws because they represent strangulation and choking hazards. Keep the kit parts out of the reach of small children.

- PicoTurbine generates low levels of electricity (1.5 volts, 200 milliAmps, about 1/3 watt) that are generally considered safe and are of the same order as produced by batteries used in toys or radios. But, to avoid shock hazard never work with electricity of any level when your hands or feet are wet.

- Persons wearing pacemakers should not handle magnets.

- Do not allow magnets to "snap" together or fall together. They are brittle and may chip or break. They can also pinch your fingers or send small chips flying through the air, presenting an eye hazard.

PicoTurbine kits should not be carried onto aircraft, because strong magnets are generally not allowed on airplanes, and because the materials used in PicoTurbine look a lot like those used for illegal devices, which could cause you delays in check-in.

Building the PicoTurbine

NOTE: The images shown are from the official PicoTurbine kit. Check out **http://www.picoturbine.com/orderform.htm** to order a kit for yourself! Using the official PicoTurbine kit will save time searching for and purchasing the necessary materials for this project.

CASE STUDY: BUILD YOUR OWN WIND TURBINE

Step 1: Glue the Template Parts

Carefully cut these out and glue them to pieces of corrugated cardboard. Any normal cardboard box will do. The template marked "Rotor" with the four square magnet outlines marked "N" and "S" should be glued to a doubled-up piece of cardboard. The two pieces of the doubled cardboard are glued together as well. Set these glued pieces aside to dry while you continue on to the next step.

Templates glued to cardboard. *The rotor template on the lower right side of this figure is glued to a double thickness of cardboard made from two glued pieces about 4 to 5 inches square.*

Step 2: The Axle and Yoke

There is a common axle used by both the blade assembly and the alternator, made from a wooden dowel. The dowel is sharpened on one end using a pencil sharpener. The dowel's point rests in the center groove of a Phillips head screw, and the blunt end is held by a wire loop. See the figure below. To make the base and yoke assembly, start with the heavy, 2-foot section of wire. Using pliers, bend a small loop on one end. Bend the loop so it forms a 90-degree angle with the rest of the wire. Measure 6 inches up from the loop and make a 90-degree bend in the wire. Measure 3 inches from this bend and form another loop, slightly larger than the diameter of the dowel. Measure 3 inches from the center of this loop and make another 90-degree bend, forming a large, square, U shape with the wire. Measure 6 inches from this bend, and form another loop. Clip off any excess wire. The U shaped piece of wire will be called the "yoke."

Fasten the yoke to the wooden base using two screws. The legs of the wire yoke should be centered on the wide face of the wood as shown above. Insert the dowel

CASE STUDY: BUILD YOUR OWN WIND TURBINE

in the center hole of the yoke and rest the point in the center screw's groove. The dowel should stand as near vertical as possible. Adjust the yoke by bending the wire if necessary to make the dowel vertical both side to side and front to back. Make sure the dowel turns freely in the yoke's center loop. If you wish, you can put a drop of any type of oil in the center screw's groove to make the dowel turn more freely.

Step 3: Cut Out Parts

Cut out the Blade Coverings from the templates. The blade may be colored or decorated using crayons or markers at this time. If you are constructing this in a mixed group then this is a good task for younger children. The ends of the blade coverings should be carefully cut into a "feathered" edge. See the figure below.

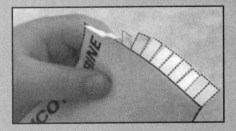

Feathered Edges. *Cutting carefully on the dotted lines will create a feathered edge that is more easily assembled.*

If the previously glued templates are dry enough carefully cut them out. The complete set of parts you need for further assembly is shown below.

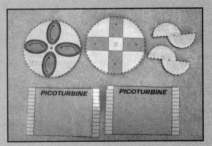

Complete set of parts. *The items on the top row are glued to cardboard.*

Step 4: The Alternator

An alternator is little more than magnets moving relative to wire loops. The magnetic

CASE STUDY: BUILD YOUR OWN WIND TURBINE

flux density changes as the magnets (or wire) move around, inducing an electric current in the wire. In PicoTurbine, the magnets will spin on an assembly called the rotor, while the wires will remain motionless on a part called the stator.

The alternator is by far the most challenging part of PicoTurbine to build. If you build it carefully, you can achieve about 2 to 2.5 volts of electricity at about 30 milliamps in a 20 mile per hour wind. This is enough electricity to light up the small incandescent lamp and the bicolor LED provided in the kit.

Step 4A – The Permanent Magnet Rotor

Tape the four magnets as on the rotor template as shown below. Note that the magnets are magnetized on their faces, and you must alternate poles going around the diameter. Remember that like poles repel, and opposite poles attract. If building this project with children, your best bet is to mark the poles using a pencil or marker before beginning. Magnets distributed with your kit are already marked with a dot on one pole. So, you should alternate: dot, no dot, dot, no dot.

Poke the dowel through the rotor as shown, being careful not to break the point. Work it down slowly so as not to stretch the hole bigger than needed, it must be quite tight. Use some tape to make sure there is a tight fit.

You can use the two rotor support tabs shown on the template to make a stronger connection. Simply cut two small slots next to the center hole of the rotor and insert the thinner side of the tab up from the magnet side. The slots should be narrower than the bulbed portion of the support tab. It is easy to affix tape to the top of the tabs to prevent the rotor from sliding down the dowel.

Place the dowel top up from the bottom into the yoke loop, pull it through, and lower

CASE STUDY: BUILD YOUR OWN WIND TURBINE

the point into the center screw. Spin the rotor by twisting the blunt end between thumb and forefinger. It should spin freely, and vertically. Adjust the wire yoke if necessary. Watch the rotor as it spins. It should spin evenly, with as little wobble as possible. Adjust it and use tape if necessary to fix it in place. If you give it a good spin, the rotor should spin on its own for quite a long time, 30 seconds or more. The ballpoint is an excellent bearing and there is very little friction. The yoke loop should not be too tight around the dowel.

Step 4B – Winding The Wire Loop Stator

If you have the PicoTurbine kit, then the wire loops are already wound for you and you can skip this section. Otherwise, take a piece of cardboard 1.5 inches wide and six inches long, and fold it, resulting in a piece about 1.5 inches by 3 inches and double the normal thickness. Tape this together so it holds. This is your wire-wrapping tool. Take your supply of 28 AWG wire. Reserve about four inches start wrapping loops around the 1.5 inch dimension. Make 300 turns of wire around your cardboard wrapping tool. Leave four inches after the last turn. Then, carefully slide the wire off the tool and immediately wrap tape tightly around the bundle of wire so it doesn't spring apart. The tighter you can form the bundle, the better. You will have a slightly oblong coil of wire, about 2 inches long and about 1 inch wide. Do this four times, creating four coils.

Test each loop to ensure it functions. Strip about an inch of wire from each end, using the blade side of the scissors or sandpaper. If you have a multimeter, set the it for AC millivolts. Holding the loop close under the magnet section of the axle/rotor assembly, give the rotor a good spin. If you spun hard and are holding the loop close to the magnets you should see 400 to 600 millivolts from a single coil.

Step 4C: Constructing the Stator

Strip the ends of the wires coming out of the coils. Stripping is best performed with fine to medium sandpaper, but you can also carefully use a knife or a scissors blade. Make sure the stripped wire is shiny copper, with no red enamel remaining. Affix the coils to the stator template as shown by the coil drawings. Note that the loops should alternate between clockwise and counter-clockwise rotation. If you are using the PicoTurbine kit, this means the part of the loop where the leads come out should alternate being near the edge of the cardboard and being near the center. Tightly twist together the stripped wires from one coil to the next, leaving the final two wires (the first and last) unattached.

Attach the coils using tape. They should lie very flat. Cut a circle in the center of the

CASE STUDY: BUILD YOUR OWN WIND TURBINE

stator cardboard. Remove the rotor/axle assembly by pulling up on the blunt end and angling it out. Put the stator assembly over the center screw, and tape it down firmly.

It will slightly overhang the ends of the wood in front and back. Put the rotor/axle back on. There should be as little gap between the coils and magnets as possible, but not so little that there is any chance of the support tab rotor cardboard, side view magnets crashing into the coils when you spin them. Adjust the center screw to adjust the height of the rotor magnets over the wire.

Now, hook the two remaining wires to your multimeter and give the rotor a spin. If you spin fast, and everything is aligned well, you should get about 1.2 to 1.5 volts (or more if you've built very well).

Step 5 – The Blade Assembly

You're almost finished! This is easy compared to the alternator. Cut the two blade supports out, and poke an X in their centers. Slide them onto the dowel from the blunt side. They should be aligned with each other, don't turn one upside down accidentally.

Glue each paper blade covering on the circular side of the blade support, both top and bottom. Use the feathered edges to negotiate around the circular support. The final effect is as if you took a cylinder, cut it lengthwise, and offset the two halves horizontally before fastening them back together.

Put tape along the two leading edges, and tape over the glued top and bottom parts just for good luck in high winds.

Step 6 – Testing

Carefully insert the blade/rotor/axle assembly back into the yoke. Blow into the blades from any direction, and they should start up very easily. Short, puffing blows are best. Hook up your multimeter again and blow in some wind. If you have very good lungs you'll get a couple of hundred millivolts, the wind will do much better than you! For classroom demonstrations a small fan or hair blow dryer can provide the wind.

Finally, if it's a windy day give it a real test using Mother Nature. Attach the mini-lamp to the leads and carefully twist them tight. In a wind of about 10 to 15 miles per hour the lamp will glow weakly, in a 15 to 20 mile per hour wind it will glow quite brightly.

CASE STUDY: BUILD YOUR OWN WIND TURBINE

Now try the LED. This is a bicolor LED. When current flows in one direction, it will glow green. In the other direction, it will glow red. Because PicoTurbine creates alternating current, it will go from green to red and back again many times per second. PicoTurbine needs to produce a minimum of about 1.5 volts to start the LED glowing, somewhat more than is required to produce a weak glow from the incandescent lamp. To produce this much power, it must turn about 3 to 4 cycles per second. A good spin with your fingers will produce this rate of turning, or a hair dryer positioned very close to the blades, or a wind of about 15 miles per hour.

Troubleshooting Your PicoTurbine

This section discusses some common problems and how to fix them. Look through this section and try out everything suggested. If you still cannot get the kit to work, send electronic mail to support@picoturbine.com and we'll give you a hand. This section discusses the most common problems.

Problem: Blades Do Not Spin, or Only Spin Slowly

1. **Yoke Too Tight.** Make sure the top of the wire yoke loop is loose enough. The dowel should be able to move slightly left and right, and should be able to turn freely. If not, use pliers to form a larger loop.

2. **Rotor/Stator Collision.** Make sure the magnets are not hitting the coils or a stray piece of tape or wire lead. If a piece of tape has come loose, clip it off or use more tape to hold it down. If the magnets are hitting the coils, move them slightly farther away and affix tightly with tape. If the magnets tend to sag to one side, then you may have to add another 4 inch cardboard disk glued to the top of the rotor to reinforce it.

3. **Bad Point.** Make sure the dowel point is reasonably sharp. If the point keeps breaking, try using a pen instead. A pen never has this problem, but most pens are a little short of the required length. You may have to shorten the blade height to accommodate a pen, unless you can find one that is a bit longer than usual.

4. **Ribbed Pen.** If you are using a pen that has "flat" edges, this can sometimes cause too much friction on the upper yoke loop. Try using a perfectly round pen instead of one that is flat.

CASE STUDY: BUILD YOUR OWN WIND TURBINE

Problem: Lamp Lights but LED Does Not

The lamp will light with somewhat less voltage than the LED. If the lamp lights weakly but the LED will not light up, then you have just barely enough voltage to light the lamp. The most likely causes of this are:

- The magnets are too far away from the coils to produce enough voltage, or

- There is friction that is causing the PicoTurbine to spin too slowly, or

- The coil connections are not tight enough are and causing too much resistance in the circuit.

Problem: Neither Lamp nor LED Light Up

1. **Lamp Burned Out.** Make sure the lamp is working by touching its ends to opposite sides of a 1.5 volt AA battery. It should glow nicely. Be sure the battery is new or test it in some device.

2. **Coils Lack Continuity.** Make sure the coils are connected well. If you have a multimeter that can test continuity or has an Ohms test, then hook one end of each alternator lead to the multimeter. A continuity check should pass. A resistance check should show approximately 32 ohms for the four coils, or about 8 ohms for a single coil. If the resistance check shows more than 40 ohms, or shows infinite resistance, then your coils are not properly connected. If you do not have a multimeter, check continuity by using the supplied lamp and a 1.5 volt AA battery. Connect the coil leads to the battery and lamp in series.

The lamp should light (assuming the lamp is good) and the battery is known to be good). The most common cause of coil problems is that the wire leads are not properly stripped. Strip each lead using sandpaper, or carefully use the edge of a knife or scissors edge. Strip about 1 inch. The stripped part should look like shiny copper with no red enamel coating at all. Tightly twist together the leads from the coils and check continuity again using a multimeter or the lamp with an AA battery. If you still cannot get a positive continuity check using multimeter or lamp, then there is a possibility that a coil is broken or kinked.

CASE STUDY: BUILD YOUR OWN WIND TURBINE

To test this, disconnect all the coils from each other and test continuity on each one individually. If you find that one coil is bad, then double check that the ends are properly stripped. Inspect the coil visually and look for damage such as kinks or breaks in the wire. If you locate a break, then strip the ends of the two broken pieces and tightly twist them together, then put a little tape around the twisted pieces.

3. **Coils Not Placed Properly.** Double-check that the coils are properly aligned. The coils must alternate direction, otherwise they will cancel each other out. Look carefully at the stator template diagram, and make sure the leads coming out of the taped portions alternate between pointing toward the center and pointing toward the radius, and make sure the wires are connected as shown.

4. **Magnets Not Placed Properly.** The magnets must also alternate going around the rotor. A dot is imprinted on one side of each magnet. The magnets should alternate going around: dot, no dot, dot, no dot. Another way of saying this is that the magnets with dot side up will be directly opposite one another, and the magnets with dot side down will be across from each other as well. If you don't do this, then power from the alternator may not be high enough to light the lamp and definitely will not light the LED.

If You Still Have Trouble

If you still cannot get PicoTurbine to work after following all of the above suggestions, you may have a defective or damaged part. Send email to support@picoturbine.com describing the problem and we'll help you out.

Teacher's Guide to PicoTurbine

Interested in having a group of young kids construct a PicoTurbine? Great! But make sure to follow PicoTurbine's advice as detailed below.

PicoTurbine makes an excellent small group project for grades 5 through High School. For the younger grades (5 and 6), it is recommended that the teacher perform some or all of the following parts of the construction in advance of the class in order to save time:

• If you are not using the kit, wind the four coils of wire in advance. Younger

CASE STUDY: BUILD YOUR OWN WIND TURBINE

children may tangle the wire and possibly break it.

- Bend the yoke wire into the U shape. Children below about grade 7 or 8 may not be able to do this accurately enough. It is also recommended for grades 5 and 6 that the teacher handle the screwdriver to screw the yoke to the wooden board and install the center screw. By performing this as a small group project, it is possible to complete the project in about 30 to 40 minutes, however to do this you must use a quick-drying glue for gluing the template parts to the cardboard.

However, for safety reasons we do not recommend "super glue" (cyanoacrylate glue) because it could bind fingers and eyelids closed if used improperly. Alternatively, you could use glue and also tape the edges after gluing. In this way, if the glue is still slightly wet later in the project you can still proceed. Alternatively, you could glue the parts in advance, or break the project across several periods (do the gluing and yoke in a 20 to 25 minute session in the morning, and complete the project in a 20 to 25 minute session in the afternoon when everything is dry). Of course, the project could also be broken across two days.

A reasonable set of tasks that can be done simultaneously by a group would be:

- One person cuts out templates and glues them to cardboard.

- Simultaneously, another group member bends the yoke wire and prepares the base screws.

- Simultaneously, if you are not using the kit, another group member winds the four wire coils.

Two group members could be assigned this task if there are enough people.

- After the above tasks are complete, one member attaches the blade coverings to the blades while another attaches the magnets to the rotor and a third attaches the wires to the stator.

- Finally, the project is assembled by the group.

Upgrading Your PicoTurbine

You can't improve upon perfection. Luckily, nothing is perfect! Here are a few ways to

CASE STUDY: BUILD YOUR OWN WIND TURBINE

keep your PicoTurbine up-to-date and in good shape.

Weatherproof PicoTurbine

PicoTurbine, as described in the main plans, is not weatherproof. The tape, glue, paper, and cardboard will quickly disintegrate in rain. Here are some ideas to produce a weatherproof PicoTurbine that can be left outdoors. It will be harder to build, but worth the effort.

- Instead of paper, use a large plastic bag for the blade covering. An alternative is to use a material such as Tyvek ™ for the blade covering, which can be purchased at hobby stores, or plastic materials used for kites.

- Instead of cardboard, use 1/8 inch plywood or corrugated plastic for the blade supports, rotor and stator. Plywood must be cut with a coping saw or keyhole saw. Corrugated plastic may be purchased at art supply or sign supply stores and can be cut with a razor blade or sharp knife. Only adults should do the cutting!

- Instead of tape and glue, use epoxy glue or weatherproof tape. These can be purchased at hardware stores. Epoxy glue should only be used by adults.

- Use small wire nuts to secure the lamp to the leads so the rain does not touch the contacts.

High Power PicoTurbine

To make a higher power version of PicoTurbine, follow these instructions:

- Make the Blade coverings 6 inches tall instead of 4 inches.

- Use a 1 foot long ¼ inch threaded rod instead of a wooden dowel. Obtain 6 nuts and 6 large washers to use to attach the rotor and blade supports.

- Use 6 magnets instead of 4. Equally space the magnets and remember to alternate north and south poles.

- Attach the magnets to a 1 gallon paint can lid. This provides a metal backing

CASE STUDY: BUILD YOUR OWN WIND TURBINE

to the magnets and increases the magnetic field strength somewhat. Use double sided tape or epoxy glue to affix the magnets. If you have magnets with holes in them you could also use screws.

- Use 6 coils instead of 4.

- This version produces about 1 full watt. Warning: the mini-lamp can't handle the voltage that will be produced. Obtain a 3-volt flashlight lamp or put 2 lamps in series. To use the LED, you must put a small resistor in series with the LED to limit the current, otherwise you will burn it out as well. Use about a 200 to 400 ohm resistor.

Alternative Blade Designs

As shown in this document, PicoTurbine uses a traditional "barrel offset" blade design. But, blades can be offset more or less. Also, the curved portion can be a shallower or deeper curve. Play around with the shape of the blade support parts and test these to see which is more efficient. This would make an excellent science fair project. It would even be possible for an advanced student to look up patented designs for Savonius wind turbines (that's the kind PicoTurbine is) and do a study of which one is best. To do this, go to the website: **http://patents.ibm.com** and search for "Savonius." You will get quite a few patents back.

Look at the blade design described in the 1996 patent by Benesh. It claims to be much more efficient than the one used in PicoTurbine. Put it to the test! Note: it is not illegal to build a model of a patent for personal testing purposes, it is however illegal to use itcommercially without permission. So, you can build the design described in this patent, but you cannot sell it to anyone!

CASE STUDY: BUILDING A DC KIT FOR YOUR PICOTURBINE

NOTE: If you haven't built the PicoTurbine documented in "Case Study: Build Your Own Wind Turbine," please do so before beginning this project.

About This Project; Necessary Electricity Fundamentals

This document will show you how to build circuits that will take the output of a PicoTurbine wind turbine and convert it into a DC voltage. These instructions assume

CASE STUDY: BUILDING A DC KIT FOR YOUR PICOTURBINE

you already have built a PicoTurbine miniature windmill (see picture below).

There are four different circuits explained in this document for converting PicoTurbine's AC output to DC. Each one requires only 10 to 15 minutes to create using a solderless breadboard which is included in the kit. To order a kit, head to **http://www.picoturbine.com/orderform.htm**.

PicoTurbine: *normally produces alternating current, but with this add-on kit can be made to produce direct current.*

A NOTE ON PICOTURBINE VERSIONS

The version 1.0 series of PicoTurbine produces about 1 to 1.5 volts at about 200 mA, while the version 1.1 series produces about 2 to 2.5 volts at about 30 mA. These experiments should work for both versions, but the text below is written such that it refers to voltages produced by the 1.1 version of PicoTurbine.

- **Voltage and Current**

Students should understand the difference between voltage and current. The simplest way is to use a water pipe analogy. Voltage is similar to the pressure in the pipe. There can be pressure, even with no water flowing (perhaps a valve is turned off). Current is similar to the volume of water flowing through the pipe. It is actually the amount of electric charge flowing through the circuit per unit time, or in other words, the rate of charge flow.

- **Resistance**

Resistance impedes the flow of electricity. In our water pipe analogy, it might be the

CASE STUDY: BUILDING A DC KIT FOR YOUR PICOTURBINE

friction of the water along the pipe walls, or perhaps the pipe is flowing uphill. Voltage (pressure) is required to overcome resistance.

- **Ohm's Law**

The relationship between Voltage, Current, and Resistance is given by Ohm's law:

$V = IR$. In other words, voltage is current times resistance. The water pipe analogy also works well here. Thinner wire has higher resistance than thicker wire. Similarly, to put a given amount of water through a larger diameter pipe requires less pressure than if the pipe is very thin. Imagine trying to pump 1 gallon of water per second through a drinking straw. That would require a great deal of pressure. However, very little pressure would be required to pump 1 gallon per second through a six-inch diameter drainage pipe. Similarly, if you want to drive a large current through a high resistance, you must have a proportionally large voltage.

- **Functions of Basic Electronic Components**

Students should understand the functions performed by diodes and capacitors. In our water pipe analogy, a diode is like a one-way valve. A capacitor is a small storage container that can store water and later release it.

- **Alternating Versus Direct Current**

Students should understand the difference between alternating and direct current. In the pipe analogy again, alternating current would be similar to water flowing in one direction, then stopping, then flowing in the other direction again. Direct current represents the more familiar situation where the "water" only flows in a single direction. It should be noted that in alternating current, both the voltage as well as the current (pressure as well as amount of water) are both varying over time.

Gathering Materials and Reviewing Safety Rules

Step 1: Check Your Materials

The following materials are supplied with your PicoTurbine-DC kit. If you did not purchase a kit, make sure to gather everything listed below:

- A small solderless breadboard and several pieces of 22 gauge prestripped hookup wire.

CASE STUDY: BUILDING A DC KIT FOR YOUR PICOTURBINE

- Four germanium diodes rated at 100 milliAmps or more continuous and peak inverse voltage (PIV) of at least 6 volts.

- Two electrolytic capacitors rated 47 microFarads and at least 6 volts.

- One piezo-electric buzzer rated at 2-12 volts DC, not requiring an external driver circuit. Not all piezo buzzers will work, the one used here is Radio Shack part 273065.

- One bicolor LED with two leads.

It is also helpful to have the following tools, but not entirely necessary:

- A digital multimeter that can measure AC/DC millivolts is useful for displaying the exact voltage created.

- A small set of needle nose pliers can make inserting and removing jumper wires from the breadboard much easier.

- If you have a really well equipped lab and happen to have an oscilloscope, this is a great way to see the actual voltage waves. Oscilloscopes are quite expensive, however, so this kit was designed so you can "hear" the waves using a piezo buzzer placed in different ways. Still, nothing is quite like looking at the patterns on a 'scope if you happen to have one or can borrow one from an electronics professional or enthusiast.

PicoTurbine-DC is not a dangerous project to build, but as with any construction project certain safety rules must be followed. Most of these rules are just plain common sense. Be sure to review these rules with children if you are building this project as part of an educational curriculum.

- The electronic components used in this kit are only capable of handling the small amounts of current and voltage produced by PicoTurbine.

UNDER NO CIRCUMSTANCES SHOULD AC CURRENT FROM A WALL OUTLET BE ATTACHED TO ANY OF THESE PARTS. HIGH VOLTAGES AND CURRENTS FOUND IN STANDARD WALL OUTLETS WILL CAUSE EXPLOSIONS, BURNS, FIRE, AND SHOCK HAZARDS, INCLUDING POSSIBLE DEATH.

CASE STUDY: BUILDING A DC KIT FOR YOUR PICOTURBINE

- Adult supervision is required for this project.

- This project is not recommended for children under 11 years old.

- Children must be supervised when working with scissors and other sharp parts to avoid cutting injuries.

- Children under 4 years old should never work with wire or small parts because they represent strangulation and choking hazards. Keep the kit parts out of the reach of small children.

- PicoTurbine generates low levels of electricity (under 3 volts) that are generally considered safe and are of the same order as produced by batteries used in toys or radios. But, to avoid shock hazard never work with electricity of any level when your hands or feet are wet.

- Persons wearing pacemakers should not handle magnets such as those found in the PicoTurbine alternator.

Inspiration for This Project

PicoTurbine produces electricity using an alternator. The word alternator is used because such a device produces alternating current, or AC for short. Alternating current constantly varies between a positive and negative voltage, usually in a sine wave pattern.

Alternating current can be used directly by many devices, such as heating elements for an electric stove or an electric light bulb. However, many devices require direct current (DC), which does not vary but remains at a steady voltage. Luckily, it is possible to convert AC to DC. The process of converting AC to DC is called rectification, and there are several different ways to do it.

To understand how to build these circuits, a little background on how the parts are named and what they do is in order.

Diode: A diode (also called a rectifier) only allows current to go in one direction. Current flowing in the opposite direction is stopped. Think of a diode as a one way valve for electricity. Diodes are rated as to how much current they can handle, and the maximum amount of reverse voltage they can withstand (this is called the Peak Inverse Voltage, or PIV).

CASE STUDY: BUILDING A DC KIT FOR YOUR PICOTURBINE

The black band around one end of the diode shows the positive voltage side.

In a circuit diagram the convention is that electricity flows from positive to negative. So, electricity could flow from the unmarked to the marked direction, while it would be stopped from the marked to the unmarked direction. The symbol for a diode shows this as an arrow.

Capacitor: A capacitor can store electrical charge. Think of a capacitor as a kind of very small rechargeable battery that can only hold a small amount of charge. Some diodes are polarized such that one side should always be positive, others are not polarized. The symbol for a polarized capacitor shows a + sign on the positive side.

The actual capacitor typically has a stripe or a minus sign on the side of the negative lead.

Light Emitting Diode (LED): An LED is a diode, and only allows electricity to pass in one direction. The symbol for an LED is the same as for the regular diode, but two small arrows indicate that light is emitted. The actual LED has one short lead and one longer lead. The shorter lead is the negative side. When current flows in the correct direction, this has the side effect of producing light. LEDs are very efficient at converting electricity to light and can operate at very low currents compared to equivalent incandescent lamp light.

The LED included in the kit is a special bi-directional LED. It is actually two LED's connected back-to-back such that when electricity flows in one direction it lights up red and in the other direction green. In a circuit diagram this is usually shown as two diodes in parallel, perhaps with labels indicating the color of each one.

Piezo Buzzer: A piezo buzzer uses the piezo electric effect to convert electricity into motion of a thin slice of crystal material. With enough voltage, the piezo buzzer will make an audible sound. The experiments in this plan use a piezo buzzer because it makes it easy to "hear" that a current is alternating or not, and even how high the voltage is (by how high the frequency sounds).

There are various kinds of piezo buzzers. Some require external circuitry to drive them. The one included in your kit requires no external driver and can take a wide range of DC voltages. Even though it is rated at 12 volts, it will produce a chirping sound that is quite easy to hear even at the 2 volt level of the little PicoTurbine. The

CASE STUDY: BUILDING A DC KIT FOR YOUR PICOTURBINE

leads are marked on the bottom as to which is positive and which negative.

Solderless Board:

The solderless breadboard allows you to quickly set up circuits without having to make solder connections.

This is faster and much safer in a classroom setting. The breadboard is marked so that rows have a number and columns have a letter. This kit will tell you to make connections using a combination of a letter and number. For example, "place a diode from A3 to A5." This would mean place the unmarked side of the diode in A3 on the breadboard and the marked side in the A5 hole.

As you can see from the picture above, there is a row marked "X" at the top and another marked "Y" at the bottom. All of the "X" holes are connected to each other. All of the "Y" holes are connected to each other as well. These are called "bus bars" and they are usually used to carry the negative (or ground) and positive voltages from the battery. They run along the length of the whole breadboard to make it easy to make ground and positive voltage connections wherever necessary.

The other rows work a little differently. Each column is divided into the upper letters (A through E) and the lower letters (F through J). Each column is connected in two sections, one for the upper and one for the lower. For example, A1, B1, C1, D1, and E1 are all connected to each other. Also, F1, G1, H1, I1, and J1 are connected to each other. These groupings are also true for all the other columns, column 2, column 3, etc. The reason for this layout is to make it easy to insert a DIP Integrated Circuit (IC) chip and then allowmultiple connections to each pin of the chip.

This document provides recommended layouts for each circuit, but many alternative layouts are possible. Keep in mind how the different parts of the breadboard are connected internally and try to visualize how the circuits shown in the schematic diagrams match to the breadboard layouts given. To insert components in the breadboard, simply bend the wire leads as necessary to reach the correct pin holes, then insert the pins approximately a quarter inch. Make sure the connections are inserted completely and firmly. Use caution when removing components so you do not damage them. Simply apply a stead pulling pressure until the leads detach. Sometimes the short jumper wires you will use to make connections will break off

CASE STUDY: BUILDING A DC KIT FOR YOUR PICOTURBINE

inside the breadboard hole. If this happens, you may need to use needle nose pliers to remove such broken wires. If you cannot remove such a wire, simply use alternative holes in the breadboard for the remaining experiments.

Connecting the PicoTurbine Alternator to the Breadboard Using Jumper Wires

When connecting the PicoTurbine alternator to the breadboard, it is helpful to first connect the alternator wires to two of the short jumper wires. Twist at least a full inch of stripped alternator wire to one end of a jumper, then pinch the bare jumper wire down against the insulation. Do this for each of the two alternator wires. Now, it is a simple matter to insert and remove the other end of each short jumper. The PicoTurbine alternator wires are a little too thin to make a good connection directly to the breadboard.

If you build one of these circuits and it fails to function as described in the text, follow these steps to solve the problem:

1.   **Check the connections.**

Double check the list of connections to make sure all the components are in the correct positions. It is common to forget a connection or miss the correct pin hole by one slot. The piezo buzzer is especially vulnerable to this because you have to carefully look from the side while inserting it to make sure its PC mount leads are going into the correct holes.

2.   **Check diode directions.**

Diodes only work in one direction, so it is critical that the black band on one side be in the proper pin hole. The instructions below always say which hole should receive the "banded" side of the diode.

3.   **Check piezo buzzer direction.**

The piezo buzzer will only work if its terminals are properly positioned. They are clearly marked "+" and "-." The directions below will indicate which hole receives the "+" terminal.

4.   **Make sure all connections are tight.**

CASE STUDY: BUILDING A DC KIT FOR YOUR PICOTURBINE

All connections must be nice and tight. At a minimum about ¼ inch of lead should be in the hole. You should feel the lead somewhat "snap" into the hole. A very slight pulling pressure should not bring the wire lead back out, it should take a bit of effort to remove in which you feel some friction. If none of the above suggestions work, you may have a defective part. You can in some cases check this.

You can check the piezo buzzer by simply connecting it to two 1.5 volt batteries in series (making sure the + lead goes to the + terminal of the battery). The buzzer should go off easily at 3 volts. You can check the LED by directly connecting it to your PicoTurbine and giving a good spin. It should flash red and green. You can check a diode in circuits that only require 2 diodes by replacing them with the other 2 diodes. If the circuit works with one diode and not another, you may have a defective or burned-out diode.

If all else fails, send email to support@picoturbine.com with a detailed description of the problem and we'll try to help.

CIRCUIT 1: A Simple Half Wave Rectifier

We'll start with the simplest possible rectifier, called a half-wave rectifier circuit. It is called this because it only allows one half of the sine wave (the positive side) to pass through. Although what comes out is technically direct current because it never goes negative, it is very "bumpy." This bumpy pattern is called "ripple."

The circuit uses only a single diode, which you will recall, acts like a one-way valve for electricity. Sometimes two diodes, one in each direction, are used in half wave rectifiers.

Building the Circuit

To build the circuit, start with a clean breadboard and make the following connections:

- PicoTurbine alternator leads go into B1 and I1. It does not matter which lead goes in which hole.

- A diode goes from A1 to X5. Place the "banded" (positive) side in X5.

- Place a jumper wire from J1 to Y5.

CASE STUDY: BUILDING A DC KIT FOR YOUR PICOTURBINE

- Place a jumper wire from X10 to A10.

- Place a jumper wire from Y10 to J10.

- Place the Piezo buzzer from D10 (positive) to H10 (negative).

Experiments With the Half Wave Rectifier

After following these steps, the X and Y rows will have the DC rectified current. Spin the PicoTurbine blades by hand, and you should hear the buzzer chirp. Now, reverse the buzzer so that it sits from H10 (positive) to D10 (negative). Spin PicoTurbine, and you should hear nothing from the buzzer. This is because current is only flowing in one direction, and the buzzer needs the current to be in the correct direction or it does not function.

Now, move the buzzer to the position D1 to H1 and spin the PicoTurbine again. This time you should hear the buzzer chirp no matter which way it is positioned. This is because the PicoTurbine is producing AC current that shifts back and forth from one direction to the other. Whenever the direction is the same as that required by the buzzer, it chirps for a moment. Now try using the LED. Remove the buzzer, and insert the LED from H10 to D10. It does not matter which lead goes in which hole. Spin the PicoTurbine and you will only see a single color light up, either red or green depending on which direction you hooked up the LED. However, if you place the LED in the position D1 to H1, it will alternate between red and green. This proves the current is running in both directions coming out of the PicoTurbine alternator, but in our rectification circuit only one direction of current occurs.

CIRCUIT 2: Smoothing DC Ripple Using Capacitors

The half-wave rectifier circuit is really too varied to be used effectively by most devices that require DC electricity. This circuit will smooth out this rough DC current using a capacitor. Recall that a capacitor acts like a miniature rechargeable battery that only holds a very small amount of charge. If we place a capacitor across the DC output lines, it will tend to take in some of the charge as the DC voltage rises, then as the DC voltage falls the capacitor will emit some of its charge. The result is that the output is considerably smoother. There is still some "ripple" in the current. The amount of ripple depends on the size (capacity) of the capacitor. Sometimes two or more capacitors of different sizes are used to smooth out ripple to within a tight tolerance that may be required by some devices.

Building the Circuit

To build the circuit, start with a clean breadboard and make the following

CASE STUDY: BUILDING A DC KIT FOR YOUR PICOTURBINE

connections:

- PicoTurbine alternator leads go into B1 and I1. It does not matter which lead goes in which hole.

- Place a diode from A1 to A5 (banded).

- Place a jumper wire from J1 to J5.

- Place a jumper wire from I5 to I15.

- Place another jumper from B5 to B15.

- Place a capacitor from E5 (positive) to F5.

- Place the buzzer from D15 to H15 (positive).

Experiments With the Smoothed Half Wave Rectifier

Spin the PicoTurbine and listen to the sound produced by the buzzer. It should sound much smoother than in the previous experiment. If you don't recall, just remove the capacitor and spin again. The circuit with no capacitor should sound a lot more varied than the one with the capacitor.

CIRCUIT 3: A Full Wave Rectifier

Half wave rectifier circuits are in effect "throwing away" half of the current by only allowing current to flow in one direction. This is undesirable because it both limits the amount of current that can be drawn from the alternator artificially, and also because it causes much more ripple in the output signal. A full wave rectifier is more complex but can overcome these problems to some degree. Although even a full wave rectifier will still have some ripple, it will be much less than the half wave rectifiers previously built.

By adding smoothing capacitors as before, this ripple can be reduced even further.

Building the Circuit

We will first build the circuit with no smoothing capacitors. To build the circuit, start with a clean breadboard and make the following connections:

- PicoTurbine alternator leads go into A5 and J5. It does not matter which lead goes in which hole.

CASE STUDY: BUILDING A DC KIT FOR YOUR PICOTURBINE

- Place a diode from D1 to B5 (banded).

- Place a diode from E1 to I5 (banded).

- Place a diode from C5 to F10 (banded).

- Place a diode from H5 to G10 (banded).

- Place a long jumper from A1 to A15.

- Place a jumper from J10 to J15.

- Place the buzzer from D15 to H15 (positive).

Experiments with the Full Wave Rectifier

Spin the PicoTurbine blades. Without smoothing capacitors it should sound "faster" than the half wave circuit. This is because each side of the wave is contributing to the sound, while in the half wave case only one side of the wave is being used. To test this, temporarily disconnect one lead of the C5-F10 capacitor, which effectively makes this circuit a half wave. Spin PicoTurbine, then reconnect the lead and spin again. You can even disconnect the lead while PicoTurbine is spinning to hear the difference. At this point you should know enough to connect a smoothing capacitor into this circuit. You will need to use some wire jumpers to do this. See how the sound compares with and without such a smoothing capacitor.

CIRCUIT 4: A Voltage Doubling Rectifier

The final circuit we will discuss is called a "voltage doubler." It allows you to rectify the current and at the same time double the voltage. Note that there is no net gain of power because the current will be cut in half, and power is voltage times current. It is possible to use additional diodes and capacitors to create voltage triplers, quadruplers, etc. In fact, some circuits used many diodes and capacitors in a sort of ladder configuration to produce very high voltages (at correspondingly very low currents) for the purposes of driving camera flashes or other purposes.

Here's how it works. Imagine that the alternator is generating positive voltage to the upper input lead in the diagram. The current will flow through the diode D1 and there will branch out, partially contributing to the positive voltage at the + DC output lead, and partially charging capacitor C1. Now consider what happens when positive voltage enters on the lower input lead from the alternator. This voltage charges

CASE STUDY: BUILDING A DC KIT FOR YOUR PICOTURBINE

capacitor C2.

Capacitors C1 and C2 will, therefore, accumulate a voltage differential of approximately the peak voltage each. They are connected in series, and so will supply a total voltage twice as high as the peak voltage.

Building the Circuit

Follow these steps to build the doubling rectifier from an empty breadboard:

- PicoTurbine alternator leads go into A1 and A10. It does not matter which lead goes in which hole.

- Place a diode from D1 to C5 (banded).

- Place a diode from E1 (banded) to G5.

- Place a capacitor from D5 to D10 (negative).

- Place a capacitor from E10 to F5 (negative).

- Place a jumper wire from A5 to A15.

- Place a jumper wire from J5 to J15.

- Place the LED from D15 to G15.

Experiments with the Doubling Rectifier

First, try the bicolor LED directly connected to the PicoTurbine alternator by placing it perhaps in holes B1 and B10 (which are unused on this circuit). Spin the PicoTurbine and the LED will alternate between green and red flashes, and will only do this at the very fastest speeds (at least when spinning by hand). Now, remove the buzzer and place the LED from D15 to G15. Now, spin the PicoTurbine. First, you will only get a single color, red or green depending on how you connected the LED. Secondly, you should notice that the LED lights up at a much lower speed than it did when directly connected.

This is because an LED can only light when the voltage is about equal to its "forward voltage drop." This is nearly 2 volts. Because 2 volts is near the upper range PicoTurbine is capable of, the LED only lights at the near-top speed of the turbine. On the other hand, using a voltage doubler cuts the current current in half but doubles

CASE STUDY: BUILDING A DC KIT FOR YOUR PICOTURBINE

the voltage, allowing the LED to light at relatively low speeds. The LED requires very little current, and there is enough available even at half normal current.

The LED may appear brighter when connected directly to PicoTurbine. This is because it is flashing both red and green, which can be visually confusing. Also, it is able to draw twice the current. The voltage doubler lets you get some light at a lower turbine speed, which is an advantage on low wind days. It does potentially reduce the maximum light you can get at higher turbine speeds. A more direct measurement of this effect can be made using a digital multimeter, if you have access to one.

First, set the multimeter on AC volts. Measure the electricity coming directly off the PicoTurbine alternator. On a fast hand-spin, you should get around 2 volts. Now, connect the multimeter to the ends of the jumpers by removing them from A15 and J15 and instead connecting them to the multimeter. Now, place the multimeter on DC volts, and again spin the PicoTurbine as fast as you can using your hand. Now you should get close to 4 volts. Note that it would be possible to redesign the PicoTurbine alternator to increase the voltage (and decrease current) at lower speeds, but this would require more turns of wire in the coils and using thinner wire. Because thinner wire costs more per pound than relatively thicker wire, and more labor would be required to wind twice as many turns, it can be more economical to use a doubling circuit than to redesign the alternator.

The Differences and Necessities of Alternating and Direct Current Flows

You may wonder why we bother with two kinds of electricity rather than standardize on either alternating or direct current. This very debate spurred a rivalry in the early part of this century between two of the greatest inventors who ever lived: Nichola Tesla and Thomas Edison. Edison wanted to simply run DC current through the distribution wires to everyone's home, while Tesla argued for 60 cycle per second alternating current. Tesla won, and we still use 60 cycle alternating current today to transmit power to our homes and schools.

The reason is that alternating current has great advantages when you are transmitting current a great distance. It is a simpler matter to step up and step down alternating current. When transmitting electricity, you want to have a high voltage and a low current. This is because resistive losses are proportional to the square of the current. So, doubling voltage cuts resistive losses by a factor of 4. Long haul power lines sometimes use 50,000 volts or more. However, high voltages like this are not safe. If we actually delivered 50,000 volts through wall outlets, thousands of people would be electrocuted to death every year.

CASE STUDY: BUILDING A DC KIT FOR YOUR PICOTURBINE

So, this high voltage is transformed down to 120/240 volts before being delivered to our homes. Even at 120 volts, electricity is quite dangerous and is not to be toyed with. But at 120 volts, we at least do not have to worry about the current spontaneously jumping out of the wall outlets. This could actually occur with very high voltages, since those voltages are enough to break down the very high resistance of air itself under the right circumstances and jump over a foot to nearby objects or people. We need DC current because many devices require it to operate properly.

Rectification circuits are commonly used by devices to translate wall outlet power (or PicoTurbine power) to DC for use by computers, radios, TVs and other devices.

We will caution here, once again, that the components supplied with the PicoTurbine-DC kit are only capable of handling the low voltages and currents produced by PicoTurbine, and could cause injury or death if used with wall outlet power. Only qualified adults with specific training should ever design circuits for use with wall outlet power.

Technical Issues

This section goes more into depth on the theory of DC and AC current and rectification. Some points were simplified in the foregoing discussions to make the project accessible to first year high school students and to simplify the explanations. These notes are for the advanced student or adults who want to know more of the details. Even these notes are somewhat simplified. The student who wishes to go still further is directed to obtain an electronics handbook. Two excellent sources of this information that are easy to understand are:

- *Electronics, A Self-Teaching Guide*, by Harry Kybett

- *Building Power Supplies*, by David Lines (a RadioShack book)

Of these, the second is the more advanced.

Diode Losses

We glossed over in the main body of this document the concept of diode losses. Diodes experience a "forward voltage drop" when rectifying current. In effect, you lose a little of the "pressure" of the electricity in the diode. This lost pressure is dissipated as heat from the diode. Different types of diode have different levels of loss. Silicon diodes typically have losses around 0.7 volts. Germanium and Shotky diodes have

CASE STUDY: BUILDING A DC KIT FOR YOUR PICOTURBINE

losses of between 0.4 and 0.5 volts. The diodes delivered with the kit are Germanium diodes. When you design an electric circuit, you must make sure the diode can handle the amount of current it will rectify. The power dissipated by the forward voltage drop is easily calculated: the power dissipation in watts is equal to the voltage drop times the current. Diodes are rated as to how much continuous current they can handle before they "burn up." Yes, they can literally burn up if you exceed the rating by enough. Usually a diode can handle a surge somewhat greater than its rating for a short period of time.

Efficiency of a Rectifier

Because of diode losses and other factors, converting AC power to DC is not a 100 percent efficient operation (as nothing ever is). In fact, it is quite inefficient in the case of PicoTurbine because PicoTurbine only generates small voltages, around 2 volts. Thus, even using a germanium diode with a 0.4 volt drop, we will lose about 0.4/2.0 = 20 percent of the voltage (and also power) just from the diode losses. There will be some additional losses because capacitors have a bit of resistance and other factors come into play.

Values of Capacitor Needed for Smoothing

The actual values of capacitor required to smooth ripple to a desired value can be calculated reasonably precisely. Some sensitive electronic components need a very smooth DC current and cannot tolerate more than a certain percentage of ripple. Although going through the math involved is beyond the scope of this document, the basic idea is that you must choose capacitors large enough so that only a small portion of their current discharges before being replenished by the next voltage peak. Some of the major factors that affect his are the capacitance value of the capacitor, the amount of current being drawn off by the device, and the voltage levels of both the AC and DC sides of the circuit. Formulas for these calculations and further discussions of ripple effects are contained in the reference books mentioned above.

In general, the less ripple that can be tolerated, the larger the values of the capacitors. However, there are usually several capacitors in parallel in such a smoothing circuit, often with different values. Smaller capacitors may be placed in series to dampen high frequency noise, while larger capacitors do the "heavy lifting" of smoothing out the current.

CASE STUDY: BUILDING TURBINE BLADES FOR YOUR PICOTURBINE

NOTE: If you have not built the PicoTurbine documented in "Case Study: Build Your Own Wind Turbine," please do so before beginning this project.

This project is not mandatory, but a little something extra from the folks at **PicoTurbine.com**.

About This Project

The design allows you to build blades in 4 foot [1.2m] tall segments. Multiple segments may be attached together to attain a larger swept area. The PicoTurbine-250 alternator will require 2 segments, resulting in a blade 8 feet [2.4 meters] tall and 2 feet [600mm] wide, with a corresponding swept area of 16 square feet (1.47 square meters). You could decide to build more or fewer segments to suit the needs of generators other than the PicoTurbine-250 alternator, however there are practical considerations as the number of segments increase. See Section X: "Technical Notes" for details.

You should allow approximately 1 hour to build each segment, with some additional time to assemble all the segments into a single rotor. For assembly of four segments, allow approximately 1 hour. Thus, to build the entire blade set for a PicoTurbine-250, allow approximately 5 hours. Your actual building time may be more or less than this depending on how good your tools and skills are. For example, having built many test models, we can now build one segment in approximately 30 minutes. There is also some savings when building multiple segments in one sitting. For example, once you cut out one blade support you can use it to easily mark the wood for all the supports you need and cut them out at one time, the same goes for drilling holes and other tasks that must be repeated for each segment. This saves you time switching between tools, changing drill bits, etc.

Gathering Materials and Reviewing Safety Notes

The following materials are needed to build the PicoTurbine 250 blade set:

- An 8-foot by 4-foot piece of corrugated plastic cut into 8 sections two feet square. Corrugated plastic can be obtained from plastics supply houses or sometimes from sign supply or sign making companies. We use 2mm thick plastic, but that is usually hard to find. The most common is 4mm which will also work.

CASE STUDY: BUILDING TURBINE BLADES FOR YOUR PICOTURBINE

- An 8-foot long piece of pressure treated 4x1 wood, cut into 2 foot lengths.

- 24 angle brackets approximately 2 inches long (each side is 2 inches) and ¾ inch wide.

- A 10-foot section of 1.5 inch diameter PVC pipe. This must be cut into two sections each four feet long (there will be a couple of feet left over).

- A 10-foot piece of 1 inch diameter galvanized water pipe. This is available at any large hardware store or plumbing supply store.

- A 4-foot section of 1 inch black pipe and a floor stand (sometimes called a flange) for 1 inch pipe, and a "T" connector for a 1 inch pipe.

- Eight pipe clamps, 2 inches in diameter. These are metal strips with a screw that allows them to be tightened around a pipe.

- Approximately 100 screws. They should be about ¾ inch long and should have a rather large pan head. Number 8 screws are ideal.

- Approximately 100 washers of a size that fits the screws above.

- Four ¼ inch diameter bolts, 2 inches long.

- Approximately 10 inches of teflon tape, ½ inch in width or more. This item is optional if you have trouble finding it. We bought ours from Grainger (**www. grainger.com**).

- Approximately 25 feet of 3/8 inch plastic coated steel cable.

- A pulley rated at 100 pounds for a 3/8 inch cable.

- Four wire rope clamps.

The following tools are needed:

- An electric hand drill with a 2 inch wood cutting bore. Also, a ¼ inch metal

CASE STUDY: BUILDING TURBINE BLADES FOR YOUR PICOTURBINE

cutting bit (such as a cobalt bit), and a 1/8 inch bit for drilling pilot holes in wood.

- A jigsaw, saber saw, coping saw, or some other saw capable of cutting curves in wood.

- A screw driver or screw driver attachment for the drill.

The PicoTurbine-250 blade set is not a dangerous project to build, but as with any construction project certain safety rules must be followed. Most of these rules are just plain common sense.

- This project is experimental in nature. There may be unknown risks associated with building this project, this list of risks does not purport to be complete. This project is intended only for adults who are willing to experiment.

- The tree-mount described in this project is undesirable from a number of standpoints. The tree may suffer damage or be killed, so do not mount this project on any tree you are not willing to put at risk. Be certain to have a good footing and use a stable ladder when attaching parts. Pole mounting is preferred.

- The pole mount described in this document is experimental in nature and has not undergone extensive testing. Construct it according to the rule: "build it so it can never fall down, and place it expecting that it absolutely will fall down." Place this structure in an area where it will not damage property if it falls, and away from areas frequented by people or animals. If you are unsure of whether you are building the structure securely enough, consult a local engineer.

- Follow all manufacturer safety guidelines when working with power tools or other tools. Appropriate eye protection and footwear should be used. Avoid loose jewelry. If any recommendations in this document conflict with manufacturer recommendations, follow the manufacturer recommendations instead.

CASE STUDY: BUILDING TURBINE BLADES FOR YOUR PICOTURBINE

Building the Blades

BUILDING A BLADE SEGMENT

A blade segment consists of a piece of 1.5" PVC pipe with two wooden blade supports which support a corrugated plastic "skin."

Step 1: Building the Blade Supports

Cut out the three templates given at the end of this document. Tape them together such that sides marked "A" match up, and sides marked "B" match up. Using the template you taped together, mark a piece of 3/4" plywood. Cut out the shape using a coping saw, scroll saw, jig saw, or other saw capable of cutting curves in wood. Drill a hole 2 inches in diameter in the center of the blade support. Repeat this process to construct a total of 4 identical blade supports.

Step 2: Attaching Blade Supports to PVC Pipe

On each blade support, attach three angle brackets as shown in the diagram. Use ¾" wood screws. Now, place a pipe clamp over the 3 angle brackets, and slip the whole assembly onto the PVC pipe. Tighten the screw on the pipe clamp. Attach the four supports so that there is precisely 2 feet between the outer edges of each pair of two supports.

Step 3: Attaching the Plastic "Skin"

Take a piece of 2 foot square corrugated plastic. Going with the "grain" of the corrugations, make a 1" wide bend along the length of the plastic. A yardstick or other piece of wood will help you make a straight bend. Make additional such bends at 1" intervals for approximately a six inch area. If you are using 4mm thick corrugated plastic (which is much more commonly available than the 2mm we use) then you may need to score through one side of the plastic. To do this, place a piece of wood to be used as a guide down the length of the plastic (in the same direction as the corrugations), and score through only the top surface of the plastic, do not score all the way through. This will allow you to easily negotiate the curved blade, but will result in some loss of strength of the material.

Starting at the tip of the curved portion of the blade, drill 1/8" pilot holes in the plywood blade supports. Space these pilot holes about 2" apart. Do this on each side of the supports, as indicated in the figure below. Starting again at the tip of the curved portion of the blade support, attach the corrugated plastic using a screw and washer.

CASE STUDY: BUILDING TURBINE BLADES FOR YOUR PICOTURBINE

The washer is needed to help distribute the pressure over a wider area of plastic helping to avoid breakage. It is best to put first one screw on the top support, then one screw on the bottom support.

Adjust the spacing between the blade supports if necessary using the pipe clamps. Continue until the plastic is secured along the length of both blade supports, then repeat for the other side of the blade. Finally, repeat this entire process for the lower pair of blade supports.

MOUNTING SEGMENTS

One or more segments may be mounted on a 1" (inner diameter) pipe. Simply place the PVC pipe of each segment over the 1" pipe. Make sure that each segment is oriented the same way, i.e. so that they will all spin in the same direction. Segments may be attached to each other simply by using screws through an adjacent pair of blade supports. Drill pilot holes to make it easier.

In order to smooth out torque and enhance startup from any wind direction, it is desirable to offset each segment as shown in the figure below. In fact, it is desirable to offset each section of a single segment. Our preferred spacing would be a 30 degree angle between each subsection, making the bottom-most subsection a right-angle with respect to the top-most subsection.

If you have access to Teflon tape, it is desirable to clean the outer surface of the 1" pipe and apply several rings of tape. The tape should be oriented such that the rotation of the blades will tend to wrap it tighter rather than to rub against the outer edge of the tape and cause it to unwrap. Several rings of tape should be placed near the top and the bottom sections of PVC pipe, where the most friction occurs.

After applying the tape, if any, you should thoroughly oil or grease the 1" pipe. This results in quite low friction between the PVC and the pipe, whether or not you are using the Teflon tape. The bottom bearing, on which the weight of the blade set rests, will be the PicoTurbine-250 alternator. If you wish to test out the blades before constructing the alternator, you may use a piece of metal or large washer, greased, to rest the bottom of the PVC pipe upon. Although this will have significant friction compared to the ball bearings of the alternator, we have done weather testing in this way and have found the blades will turn quite nicely with this simple bearing except in the lowest of winds. If you have a ball bearing of the proper size, or could make it the proper size using wooden inserts, this would make for a better bearing for weather testing.

CASE STUDY: BUILDING TURBINE BLADES FOR YOUR PICOTURBINE

MOUNTING THE BLADES OUTDOORS

There are several different ways to suspend the pole. The ideas presented here are to be considered only suggestions; none of them has sufficient testing to insure they are safe and effective. Always follow the adage: "mount your wind turbine in a way that it cannot possible fall; and place the turbine expecting that it absolutely will fall." Be sure that a setting is selected that is clear from areas that people or livestock regularly traverse, and that if the structure falls no property, buildings, electric wires, etc., will be damaged.

Mounting is probably the most dangerous part of wind turbine deployment. Be sure to use a good sturdy ladder with good footings on the ground. Always plot your moves carefully to avoid falling. Never attempt this in high wind conditions or during bad weather. Use common sense.

Tree Mount

For our initial testing we used tree mounts. We should point out that tree mounts have many disadvantages, among which is the possible death of the tree involved. In our case the convenience of this type of mount outweighed the disadvantages for initial testing, however we are testing out pole mounts for actual production.

Our site has many 40 foot tall oak trees, which make reasonably good mounts. These trees are about 1 foot wide at the base of the trunk and so are quite sturdy. The branches on oak trees start quite high, about 20 feet, making for a clear section that makes mounting easy. Other types of tree will not work as well due to branches that start lower. Evergreen trees such as spruce would be completely unsuitable due to their shape.

A simple configuration using standard pipe parts is easily constructed. To prevent the pipe parts from unscrewing themselves (a problem we had with early prototypes), we drill a ¼" hole near the connectors on the pipes and use bolts to affix a small section of angle iron. To hoist up the pipe and provide added safety, we used plastic coated steel cable and a pulley. We leave the steel cable and pulley attached to the tree for easy lowering if ever necessary, and for added safety in case the mountings come loose. The bottom of the steel cable is attached to nails driven in the trunk near the base.

We secured the top section of pipe first by nailing the pipe flange to the tree and screwing in the four foot section and 90 degree angle fitting, securing with the angle

CASE STUDY: BUILDING TURBINE BLADES FOR YOUR PICOTURBINE

iron discussed above. The blade section can then be inserted from the bottom, along with any washer or steel disk used as a temporary bearing. Then, the "T" connector is attached and a second four foot section is affixed to the tree, along with another safety bracket made from angle iron. If desired, another section of pipe can be placed on the lower section of the "T" and buried in the ground. A permanent footing of concrete would be suggested for a permanent installation. In our temporary testing stations we often just use several cinder blocks or large rocks to stabilize the bottom pole, but that would not be desirable for a permanent installation.

Sometimes the pole will not be straight as mounted above because the tree trunk narrows or curves slightly as it gets higher. We have found that there is little measurable effect on performance if the pole is not perfectly straight. As long as the top is within a few inches of vertical when compared to the bottom it should be ok. You could use longer or shorter sections of pipe for the upper or lower supports to adjust if necessary, or build out the top or bottom flange from the tree by placing a piece of wood board between the flange and the tree.

Pole Mount

A pole can be used in place of a tree for a much more desirable mounting structure. We are experimenting with 4x4 inch poles 14 feet long. Such poles must be sunk several feet into the ground and guyed with wire. It would be possible to put 2 turbines on a single pole as shown in the figure below, as long as you are in an area with a "prevailing wind" this will not hurt performance significantly.

Position the two turbines such that they do not block each other during a typical "prevailing wind." Another type of multi-turbine installation that could take advantage of a prevailing wind direction would be a line of turbines supported by a horizontal wooden beam. This beam could either be suspended by other beams or pipe poles in concrete footings, or could be suspended from two trees (if you are willing to risk losing two trees). In this case the flanges could be connected directly to the mail axle pipe without the cross-bar piece. The angle bracket to prevent unscrewing of the flange would be connected from the axle pipe to the top beam in this case. Once again, guy wires would be needed to ensure stability in high winds.

In most cases, it is possible to do maintenance near the ground. This is because the alternator is at the bottom already, so all electrical connections and the most complex part of the turbine are already near the ground. If the blades ever need to be taken down, it is possible to simply disconnect the lower safety angle brackets, unscrew

CASE STUDY: BUILDING TURBINE BLADES FOR YOUR PICOTURBINE

the lower horizontal support pipe from the flange, unscrew the "T" connector, remove the bolts holding the alternator stator to the pole, and then simply lower the whole structure down from the pipe. There is rarely a need to go back up a ladder because of this ability to remove the turbine from ground level. The only regular maintenance that would require a ladder would be to grease the entire pipe, and even that could often be done simply by using a oil rag on a pole.

Alternative Construction Suggestions

We chose these materials after months of investigation, in which we tried many different materials. But, there are always choices to be made, and we will discuss some of them here.

Corrugated Plastic Characteristics

The best characteristics of the corrugated plastic material we decided on are:

- Low cost, a 4x8 foot sheet only costs about $15.

- Lightweight, a whole sheet only weighs a few pounds.

- Safe, no sharp edges.

- Easy to work with. Can be cut with ordinary razor blade.

- High strength. This material is actually used to make packing containers that hold hundreds of pounds of materials.

- Very quiet. Even in high winds this material gives off almost no sound that can be heard beyond a few dozen paces.

- Easy to obtain. This material is widely used for signs and can be found world wide in sign stores or plastics supply companies.

- Recyclable, at the end of its operating life the material can be recycled.

The only drawback we are aware of is that the material is not UV tolerant. According to the manufacturer, when used as a sign material it will start to degrade in two to

CASE STUDY: BUILDING TURBINE BLADES FOR YOUR PICOTURBINE

three years. We have been weather testing pieces of plastic for about five months now, and we have seen no deterioration yet. However, we have not had extreme cold temperatures of winter yet, so only time will answer the question of how long it will last in a windmill application.

Also, we would like to point out that just because the material does not last more than 3 years in a signage application does not necessarily rule out that it will last longer in our application. While the material will be under much more stress when used as a wind turbine blade material, the judgment of whether it is still usable is quite different from a plastic outdoor sign! In fact, the manufacturer may have meant that color fading and minor imperfections would start to occur in two to three years. In a wind turbine application such considerations are not important.

The only important thing is whether it holds together. We believe some patching could be done to extend the life the material as well. A little duct tape might repair minor defects that arise over the years and extend the life significantly. In any case, we do not feel a $15 replacement cost once every few years is a major problem. Using an electric screw driver, the blade coverings could be replaced in about one hour. This would be quite within a normal maintenance requirement of most commercial wind turbines in terms of both time and money.

Other Possible Blade Coverings

- **Sheet metal.** We rejected aluminum flashing material after testing it last winter. It was found to be very noisy in high winds. It also suffered from small fatigue cracks after only a few months, and we feel it would be quite shredded in a year or two. It is also rather sharp and presents cutting hazards. It is possible that a heavier grade of aluminum sheet metal would work properly, but it would be unlikely to be as lightweight and easy to work with as the corrugated plastic.

- **Sail cloth.** We rejected these materials based on price. There are many different grades and types, but all the ones we looked at were between five to twenty times more expensive than the plastic. We never tried to see how they worked, however, and sail cloth might be fine to use if you have a source of used or surplus material.

- **Tyvek.** This material is used to wrap houses and even to make lab coats.

CASE STUDY: BUILDING TURBINE BLADES FOR YOUR PICOTURBINE

We tried it, and it shredded to pieces in the first 40 MPH wind storm. A heavier grade might work better, though. It was also rather noisy, sounding like large sheets of paper being rattled. We did not pursue investigating other grades of the material.

- **Other sheet plastics.** There are more grades and types of sheet plastic than we can mention here. It is quite possible that a much better material is out there. In terms of price and availability, though, the corrugated plastic is hard to beat. Every other plastic material we looked at was more expensive for the same strength and size characteristics, or would be very hard for people to find locally.

- **Fiberglass.** This was both more expensive, and also heavier, than the plastic material. But if you care to make things using fiberglass it would probably make a very nice blade set. It is commonly used for horizontal axis wind turbine blades. It would probably last more or less forever at the slow speeds of a Savonius.

HIGH WIND AREA MODIFICATIONS

We are testing our designs in Northern New Jersey, USA. We are in a class 3 wind zone, rather average. If you are in a much windier place, such as a class 4 or 5 wind zone, you might consider beefing up the blades to handle the larger amount of wind. Here are some ideas we have had, but have not tested. We would enjoy hearing other ideas, or hearing whether these ideas were useful. It is important to note that these reinforcement ideas add expense to the wind turbine, and also add weight and extra inertia that detracts from startup. If you are in a high wind area then neither of these caveats are of much concern—if you have a lot of wind you need a stronger machine, and the added cost is made up by the fact that you will be getting a greater amount of energy out of the machine. In a high wind area the extra inertia of a heavier blades is also not as much of a concern, since presumably there is plenty of wind on most days for start up purposes.

Reinforced Plastic Edge

The place where the plastic meets the wood, held in place by screws, is under a good deal of stress in high winds. It is possible that over time the screws will eventually pull through the plastic. This top edge could be reinforced by placing a three-quarter inch strip of metal along the length of the plastic, and punching holes in this metal for the screws to be inserted through. Such strips can be found in hardware stores, they

CASE STUDY: BUILDING TURBINE BLADES FOR YOUR PICOTURBINE

are used in plumbing for hanging pipes in basements, and already have holes in them in some cases. They are often made from copper which will not rust. This strip would help distribute the pressure from the screws along a wider area and would also offer some weather protection and UV protection to this crucial area of the plastic covering.

Another area of concern at the wood/plastic interface is that the screws will pull out of the plywood. If a good grade of plywood is used and pilot holes are drilled for each screw the hold should be good. But, over time as the wood wears out the screws may start to pull through. Small angle brackets could be placed once every few screws to combat this. The angle brackets could be screwed in both on the edge where the plastic is, and also on the top of the blade support, through the plywood rather than edgewise. This would be most important at the curved area of plastic, and less so on the flat area that is under less stress. So, you could put several brackets on the curved portion, then space them out more along the flat section, perhaps only having two or three along the entire flat area.

The leading and trailing edges of the plastic could also be reinforced in a number of ways. A simple piece of duct tape or some other weatherproof tape would help shield this part from weather and give it more strength.

Another idea would be to thread a heavy wire through the corrugation along this edge and attach the wire to the top and bottom of the blade supports with a screw. This would serve to reinforce the entire length of the leading edge, which faces the heaviest stream of wind.

Reinforced Blade Support

The blade support can be made from 1" thick wood instead of plywood. There is a lot of waste in this case because you must use a 12" wide board and much of it will end up being cut away. Another idea is to just reinforce the central section by running a piece of 1x4" wood underneath the plywood, stopping when you get to the curved sections. This can be affixed using about a dozen screws, put in from the plywood side with pilot holes.

Doubling up the pipe clamps by using two per blade support would give an extra bit of hold to the blades against consistently strong winds. Longer angle braces might be needed in this case to accommodate the two clamps.

CASE STUDY: CAPE WIND

Renewable Energy Made Easy (REME): Would you mind telling us a bit about yourself, Mark? How did you become interested in renewable energy, and what is your involvement with the Cape Wind project?

Mark Rodgers: My title is Communications Director. I was very attracted to the project as soon as I heard about it because it was the first utility-scale renewable energy for the region that had ever been proposed. It was exactly the sort of project I was looking for. I had been involved on a volunteer basis with an organization I helped form for a number of years which was north of Boston. We were raising awareness concerning health affects from power plants; one I lived near in particular. We were advocating the greater use of renewable energy, and it was frustrating because there was so little of it around.

When I first heard about Cape Wind, in its earliest days around mid-2001, I was very excited and attended all of the early events as a third-party advocate. Over the course of a few months, I got to meet some of the principals in the company. One thing led to another, and in January of 2002, I was hired.

REME: How did the idea for the Cape Wind project come about?

Mark Rodgers: The company behind Cape Wind is Energy Management Anchor (EMI). It is a 30 year-old Massachusetts-based company that, in the twenty or thirty years before proposing Cape Wind, had been mostly involved in the field of developing and operating natural gas power plants. They did six of those projects in New England. As time passed, the company felt that they could not continue to rely solely on natural gas; there's a limited amount of it available to our region, and prices were going up.

EMI had long been interested in renewable energy. They had done a biomass power plant and was thinking that now might be the time--this was in the late '90s -- that now might be the time to really focus on the possibility of using wind power on a large scale. By 2001, that had coalesced into the Cape Wind proposal.

Cape Wind is still in the pre-construction stages.

REME: Taking in factors such as land siting, construction, and the cost of parts to build

CASE STUDY: CAPE WIND

all of the turbines, how much would you estimate for the Cape Wind project's final cost?

Mark Rodgers: The project will cost in excess of one billion dollars.

REME: How many wind turbines will be used for Cape Wind, and how much electricity will they generate?

Mark Rodgers: The proposal is for 130 wind turbines that are 3.6 megawatts each.

REME: What sort of winds will these turbines need to constantly generate electricity?

Mark Rodgers: They hit maximum power output at 27 miles per hour of wind.

REME: What other boons will the Cape Wind project provide?

Mark Rodgers: Cape wind will reduce air emissions, reduce carbon dioxide. It will provide jobs, and more stable electricity pricing.

REME: Is the goal of the Cape Wind project to reduce the cost of electricity?

Mark Rodgers: No, I don't think it's a case of reducing costs. I think it's a case of being able to provide stable, long-term costs in an otherwise very volatile price mix of fossil fuels. The price of wind power has gone up, as commodity prices have gone up. The price had been coming down, but in recent years, all new forms of energy production have gotten more expensive in recent years. That's particularly true of wind because not only have commodity costs gone up for things like steel and copper, but there's also been something of a worldwide shortage of wind turbines. Demand has been outpacing supply in recent years.

REME: Research indicates that the most common problem in the construction of a wind farm is local opposition; specifically, visual pollution. How much support has the Cape Wind project garnered?

Mark Rodgers: Independent polling data shows that in Massachusetts, support is very strong, between 71 and 84 percent, looking at a couple of different polls. The most recent polls on Cape Cod in the islands found 61 percent support. Opinion is divided, but it's moving in our direction.

REME: In 2006, a section was added to the Coast Guard re-authorization bill that

CASE STUDY: CAPE WIND

would have prevented offshore construction within 1.5 miles of any shipping channels. Oddly, one of the proponents of the bill was Edward Kennedy, a man who had verbally committed himself to renewable energy; specifically, wind power.

Mark Rodgers: Fortunately, [that incident] had a positive ending. There were some attempts at unwanted legislation having to do with Coast Guard authorization that would have placed unreasonable restrictions on offshore wind projects. Fortunately, those provisions were dropped. There was enough of an outcry among supporters of renewable energy to decrying what would have been bad energy legislation. We survived that episode.

REME: Other than the mention of Cape Wind in *Renewable Energy Made Easy*, has Cape Wind been featured in any other books or documentaries?

Mark Rodgers: The *Wind Over Water* documentary (**http://www.WindOverWater.org/**) happened pretty early on. A film student shot that. There has been interest expressed by different groups over the years in doing a different documentary.

The book was great, something the authors had been working on for a few years. The title is Cape Wind, and that was gratifying to see so much of the story told in a way that I thought was accurate, but also interesting. It was surreal, reading a book about something that I was involved in.

REME: If you had to guess, what would be your time of completion for the Cape Wind project?

Mark Rodgers: We think -- we hope -- that by the end of 2009, we will be fully permitted. We would like to be up and running in 2011.

CASE STUDY: PEANUT POWER

Thanks to Chris Graillat and the folks at the California Energy Commission (**www.EnergyQuest.ca.gov**), the true potential of the peanut can be realized: a source of fuel! Read on to learn how much power is actually contained within those rinky-dink peanut shells.

A note to younger readers: Fire is involved in this experiment, so please have an adult help you.

CASE STUDY: PEANUT POWER

Just about everything has potential energy stored in it. The problem is releasing that energy to be able to do some work.

A tiny peanut contains stored chemical energy. When we eat them, the stored energy is converted by our bodies so we can do work. We can also use the energy in a peanut to heat a container of water.

Gathering Materials

1. A small bag or can of unsalted, shelled peanuts

2. A cork

3. A needle

4. A large metal juice or coffee can

5. A small metal can (like a soup can) with paper label removed

6. A can opener

7. A hammer

8. A large nail

9. A metal BBQ skewer (like the kind for kebobs)

10. About a cup of water

11. A thermometer

12. Some matches or a lighter (ask an adult for help here)

13. A piece of paper and pencil to record your observations

Conducting the "Peanut Power" Experiment

Carefully push the eye of the needle into the smaller end of the cork. Then gently push the pointed end of the needle into a peanut. If you push too hard the peanut will break. If it does, use another peanut. It's also better to have the peanut at a slight angle. Remove the two ends of the large juice can with the can opener. Be careful

CASE STUDY: PEANUT POWER

as the top's and bottom's edge can be sharp!

Using the hammer and nail, have an adult punch holes around the bottom of the large can. These are air holes that will make the can act like a chimney and will contain the heat energy focusing it on the smaller can.

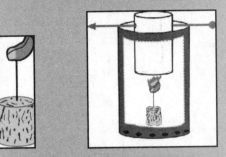

Remove the top end of the small can if it is not already removed. Using the hammer and nail, punch two holes near the top of the small can exactly opposite each other. Slide the BBQ skewer through the holes of the small can.

Pour 1/2 cup of water into the small can and let it sit for an hour. This will allow the water to be heated or cooled to room temperature. (Munch on some peanuts while you're waiting.) Put the thermometer into the water and record the temperature on your paper.

Place the cork and peanut on a nonflammable surface. Light the peanut with a match or lighter. **Have an adult help you!** Sometimes the peanut can be difficult to light, so the lighter may be easier to use.

As soon as the peanut has caught fire, immediately place the large can around the nut. Balance the skewer holding the small can on the top of the large can. Allow the nut to burn for several minutes or until it goes out. Stir the water with the thermometer and record the temperature again.

Analyzing the Results

The chemical energy stored in the peanut was released and converted into heat energy. The heat energy raised the temperature of the water in the small can.

Try a couple of other experiments using different kinds of peanuts or other kinds of nuts. Try:

CASE STUDY: PEANUT POWER

- Raw peanuts

- Dry roasted peanuts

- Vacuum-packed peanuts

- Freeze-dried peanuts

- Try cashew nuts, Brazil nuts, pecans, walnuts or other kinds of nuts. (Do they contain more energy than the peanut? Why or why not?)

You might want to try more than one peanut. You'll need extra needles. Use four or five peanuts to heat the water. Is the temperature four or five times higher?

Energy is measured in a unit called the Btu, which stands for British thermal unit. A Btu is the amount of heat required to raise the temperature of one pound of water by one degree Fahrenheit. Using math, you can figure out how many Btu are in the one peanut. (The plural or Btu is still Btu, not Btus.)

First you'll need to find out how heavy 1/2 cup of water is. Use a small scale and weigh the small can with nothing in it. Then weigh the can with 1/2 cup of water in it. That will tell you how much the water weighs.

Then, knowing how hot the water was, how many degrees its temperature was raised, you can figure out roughly how many Btu are in the peanut. (**PLEASE NOTE**: This will be an *approximate* figure because the entire peanut will not be completely burned...there is still some chemical energy left inside the partially burned peanut. In order to measure the heat energy exactly, you would need to use a sophisticated piece of machinery called a "calorimeter.")

For example: If the water weighed four ounces (1/4 of a pound), one Btu would raise the water temperature 4 degrees Fahrenheit. So, if your water temperature increased by 10 degrees (70 degrees at room temperature to 80 degrees), 10 divided by 4 would mean the peanut contained approximately 2.5 Btu. *This is only an example of the math and will not be the same as your calculations.*

One Btu equals approximately:

- One blue-tip kitchen match

- 0.252 kilogram Calories (food calories)

CASE STUDY: PEANUT POWER

1000 Btu equal approximately:

- One average candy bar (252 kilogram Calories)

- One hour of bicycling

- 4/5 of a peanut butter and jelly sandwich

NOTE: You may see Btu defined as 252 calories. These are *International Table* calories which are equal to 1000 of the "Calories" or "kilocalories" we use for measuring food energy.

GLOSSARY

Active solar: Devices or systems that use mechanical components such as fans to capture solar radiation and convert it into a usable form such as electricity.

Aerodynamic noise: The interaction between something heavy and the wind current, which causes swishing noises.

Aerofoil: A device that, when in motion, is able to provide the turbine with a reactive force that is relative to the air surrounding the turbine.

Anaerobic: Pertaining to or caused by the lack of oxygen.

Antinode: A point of max amplitude from an entire series of waves.

Attenuator: A wave energy conversion device designed based on geometry and wave orientation; sees its principal axis set perpendicular to the front of incoming waves.

B100: Pure biodiesel.

Balance-of-plants components: A variety of trinkets and

gadgets including foundations, transformers, and other trinkets, all of which play small yet vital roles in ensuring a wind turbine's operability and stability.

Balance-of-system components: Components found in photovoltaic systems. Examples: a step-up transformer to increase the voltage, an inverter used to convert the direct current (DC) produced by the module to the grid's alternating current (AC), and simple but oft-forgotten materials such as supports, miscellaneous hardware, and brackets help maintain photovoltaic modules.

Biodiesel: A diesel-equivalent fuel that can be used individually or in conjunction with conventional diesel fuel in unmodified diesel engines.

Biofuels: Fuels produced from renewable biological resources such as plant biomass. Example: methane gas.

Biogas: A gas produced by the biological breakdown of organic matter in the absence of oxygen. Biogases are primarily comprised of carbon dioxide and methane.

Biomass: Biological materials such wood scraps and sawdust leftover from lumber processing, forest and crops grown specifically for biomass, non-edible portions of crops left behind after a harvest, tree stumps, plant remains, forest residue, sewage and farm animal waste, agricultural scrapes such as pits, plant stalks, shells.

Breaking wave: A wave whose amplitude reaches a critical level at which some process can suddenly start to occur that causes large amounts of wave energy to be dissipated. Also known as a "breaker."

Caisson: Underwater structure that retains water by sealing it in, but can also pump out the water.

Co-firing: The process of adding a small amount of a biofuel to coal.

Conduction: The transfer of heat back and forth between hot and cold regions.

Conservatory: A greenhouse that is attached to the rear of a building. Heat is collected within the conservatory and subsequently transferred into the attached building. See also, "sunspace."

Convection: The process by which a fluid expands and increases in density after it has been warmed.

Crystalline silicon: See, "silicon."

Daylighting: The practice of placing windows and other reflective surfaces so that, during the day, natural light provides effective internal illumination.

Direct combustion: The process of directly burning biomass materials in a steam boiler to convert water into steam, which drives generators.

Direct gain: A device or system that uses windows to the sun's rays and heat the interior of a structure.

Direct solar: Devices or systems that make use of a single-step procedure to convert sunlight into usable energy.

Distilled: A purification process dictating that the liquid is vaporized and subsequent condensation. The distillation process requires tremendous amounts of heat, which is typically supplied by discarded crop wastes.

Doping: A process in which the properties of a semiconductor are intentionally changed due to an introduction of impurities into the semiconductor.

Earth sheltering: The architectural practice of using earth against building walls for external thermal mass, to reduce heat loss, and to easily maintain a steady indoor air temperature.

Emissivity: The measure of a material's ability to both absorb and radiate energy.

Ethanol: A flammable, colorless chemical compound, best known as the alcohol found in alcoholic beverages.

Fermentation: The process of energy production in a cell under anaerobic conditions.

Fetch: The distance over which wind energy is transferred into the sea to form waves.

Fish ladders: Structures such as steps which serve to aid fish in their transportation.

Free heat gains: Contributions made to a building's heating by way of natural activities. Examples: cooking, washing and drying clothes and dishes, lighting, using appliances, and body heat.

Fuel variability: The tendency of some fuels to be moister than others and to contain fluctuating amounts of contaminants, in addition to other variable concerns.

Gasification: The process by which a fuel is converted to a combustible gas and subsequently used to directly drive a turbine.

Gasohol: Petrol containing 26 percent ethanol.

Geothermal energy: Energy generated beneath the surface of the earth.

Gober gas: A biogas generated from cow manure; it is almost completely smokeless and odorless after being properly treated and converted.

Greenhouse effect: The process in which the emission of infrared radiation by the atmosphere warms a planet's surface.

Head: The elevation of water, divided into low, medium, and high.

Heating seasons: Times of year during which optimal opportunities arise to accomplish space heating. Heating seasons fluctuate depending on location and time.

Indirect solar: Devices or systems that make use of procedures that involve multiple steps to accomplish the same goal.

Latching: Holding the floating piece of a device under water for just about one full second before allowing the device to follow the wave.

Light tubes: Cylindrical tubes made of highly reflective materials.

Mechanical noise: A form of noise pollution caused by clinks and clanks from various pieces of machinery.

Nacelle: The housing on top of a wind turbine's tower.

Noise pollution: Displeasing human or machine created sound that disrupts the activity or happiness of human or animal life.

Passive solar: Devices or systems making use of non-mechanical techniques to capture and convert sunlight into a usable, beneficial form.

Period: The time it takes the wave to pass a certain, specific point.

Penstock: A device responsible for carrying water from the reservoir to the turbine. A penstock could be any number of devices, but is most commonly seen as a gate.

Petrol: Gasoline.

Photosynthesis: The process by which living organisms such as plants convert light energy into chemical energy.

Photovoltaics: a) A solar power technology that uses solar cells or solar photovoltaic arrays to convert light from the sun directly into electricity. b) The field of study relating to this technology; there are many research institutes devoted to work on photovoltaics.

Point absorber: A wave energy conversion device designed to draw a wave's energy from the water beyond the physical limitations of the device, but have rather small physical dimensions relative to the length of a wave.

Radiation: The process in which energy is emitted by one body, transmitted through an intervening medium or space, and absorbed by another body.

Refraction: A change of wave direction.

Runner: The rotation system of a turbine; responsible for spinning the blades.

Sea state: Comprised of three important variables: the height, the period, and the character of waves on the large body of water.

Silicon: A non-metallic element, having amorphous and crystalline forms, occurring in a combined state in minerals and rocks.

Siting: The process of choosing an ideal location for a device or structure.

Solar collector: A device designed to capture and use solar radiation for heating air or water and for producing steam to generate electricity.

Solar energy: Energy from the sun. See also, "solar power."

Solar power: Energy from the sun. See also, "solar energy."

Solar savings fractions: The amount of solar energy provided based upon dividing solar technology by the total amount of required energy.

Sorghum: A grassy type of grain crop that greatly benefits ethanol production in Brazil and the United States.

Storm waves: Waves that are located close to or directly within the area where they were originated.

Stratification: The tendency for warm air to become trapped in a skylight well. Stratification leads to increased heat loss in cool climates.

Sunspace: A greenhouse that is attached to the rear of a building. Heat is collected within the sunspace and subsequently transferred into the attached building. See also, "conservatory."

Sustainable energy source: Something that is not easily depleted by continued use.

Swell waves: Storm waves that have traveled far from their areas but have undergone low levels of energy loss.

Swept volume: The total volume of air and fuel mixture an engine can draw in during a complete cycle of a wave-capturing device's engine.

Terminator: A wave energy conversion device built so that its principal axis is parallel to the front of a wave. A terminator device is also built to physically intercept a wave.

Thermal conductivity: A measure of how well a specific material is able to transfer heat.

Thin-film: A thin material able to produce electricity.

Toroid: A doughnut-shaped object.

Tower: The body of a wind turbine.

Transmittance: A fraction of incident light that passes directly through glass.

Trellis: A construction of wood typically used to support a climbing plant.

Trombe wall: A wall facing the sun that has been constructed from thermal mass materials such as metal, stone, water tanks, concrete, or adobe.

Visual pollution: Unattractive or unnatural visual elements of a vista, a landscape, or any other thing that a person might not want to see. Example: wind turbines placed on a particularly attractive plot of land.

Volute: A scroll-shaped tube that diminishes in size while wrapping around, much like a snail shell.

Wind farm: A group of wind turbines.

Wind speed-frequency distribution: A graph that depicts the number of hours during which the wind blew at a variety of speeds for a set period of time.

Wind speed-power curve: A result which takes into account the area swept by the wind turbine's rotor, the choice of the turbine's aerofoil device, the number of blades attached to the turbines, the shape of each of the blades, the optimal blade tip speed, how fast the turbine will be able to rotate (this factor alone is dependent upon blade shape, size, and angle), the aerodynamic design of the wind turbine's blades (and, to a degree, the turbine itself), and the efficiency levels of both the gearbox and the generator.

Wind turbine: A turbine powered by the wind.

Yawing mechanisms: Machinery that quickly realigns a wind turbine's rotor whenever the wind's direction undergoes a change.

APPENDIX - ADDITIONAL TERMS

Acid rain: Also called acid precipitation or acid deposition, acid rain is precipitation containing harmful amounts of nitric and sulfuric acids formed primarily by sulfur dioxide and nitrogen oxides released into the atmosphere when fossil fuels are burned. It can be wet precipitation (rain, snow, or fog) or dry precipitation (absorbed gaseous and particulate matter, aerosol particles or dust). Acid rain has a pH below 5.6. Normal rain has a pH of about 5.6, which is slightly acidic. The term pH is a measure of acidity or alkalinity and ranges from 0 to 14. A pH measurement of 7 is regarded as neutral. Measurements below 7 indicate increased acidity, while those above indicate increased alkalinity.

Acquisition (foreign crude oil): All transfers of ownership of foreign crude oil to a firm, irrespective of the terms of that transfer. Acquisitions thus include all purchases and exchange receipts as well as any and all foreign crude acquired under reciprocal buy-sell agreements or acquired as a result of a buy-back or other preferential agreement with a host government.

Acquisition (minerals): The procurement of the legal right to explore for and produce discovered minerals, if any, within a specific area; that legal right may be obtained by mineral lease, concession, or purchase of land and mineral rights or of mineral rights alone.

Active power: The component of electric power that performs work, typically measured in kilowatts (kW) or megawatts (MW). Also known as "real power." The terms "active" or "real" are used to modify the base term "power" to differentiate it from Reactive Power.

Agriculture: An energy-consuming subsector of the industrial sector that consists of all facilities and equipment engaged in growing crops and raising animals.

Alternative fuel: Alternative fuels, for transportation applications, include the following:

* methanol

* denatured ethanol, and other alcohols

* fuel mixtures containing 85 percent or more by volume of methanol, denatured ethanol, and other alcohols with gasoline or other fuels -- natural gas

* liquefied petroleum gas (propane)

* hydrogen

* coal-derived liquid fuels

* fuels (other than alcohol) derived from biological materials (biofuels such as soy diesel fuel)

* electricity (including electricity from solar energy.)

Base load: The minimum amount of electric power delivered or required over a given period of time at a steady rate.

Bitumen: A naturally occurring viscous mixture, mainly of hydrocarbons heavier than pentane, that may contain sulphur compounds and that, in its natural occurring viscous state, is not recoverable at a commercial rate through a well.

Boiler fuel: An energy source to produce heat that is transferred to the boiler vessel in order to generate steam or hot water. Fossil fuel is the primary energy source used to produce heat for boilers.

Calcination: A process in which a material is heated to a high temperature without fusing, so that hydrates, carbonates, or other compounds are decomposed and the volatile material is expelled.

Coking: Thermal refining processes used to produce fuel gas, gasoline blendstocks, distillates, and petroleum coke from the heavier products of atomspheric and vacuum distillation.

Combustion: Chemical oxidation accompanied by the generation of light and heat.

Design head: The achieved river, pondage, or reservoir surface height (forebay elevation) that provides the water level to produce the full flow at the gate of the turbine in order to attain the manufacturer's installed nameplate rating for generation capacity.

Desulfurization: The removal of sulfur, as from molten metals, petroleum oil, or flue gases.

Diesel fuel: A fuel composed of distillates obtained in petroleum refining operation or blends of such distillates with residual oil used in motor vehicles. The boiling point and specific gravity are higher for diesel fuels than for gasoline.

Diffusive transport: The process by which particles of liquids or gases move from an area of higher concentration to an area of lower concentration.

Electric current: The flow of electric charge. The preferred unit of measure is the ampere.

Electric energy: The ability of an electric current to produce work, heat, light, or other forms of energy. It is measured in kilowatthours.

Electric power: The rate at which electric energy is transferred. Electric power is measured by capacity and is commonly expressed in megawatts (MW).

Foundry: An operation where metal castings are produced, using coke as a fuel.

Gas: A non-solid, non-liquid combustible energy source that includes natural gas, coke-oven gas, blast-furnace gas, and refinery gas.

Global warming: An increase in the near surface temperature of the Earth. Global warming has occurred in the distant past as the result of natural influences, but the term is today most often used to refer to the warming some scientists predict will occur as a result of increased anthropogenic emissions of greenhouse gases.

Gross domestic product (GDP): The total value of goods and services produced by labor and property located in the United States. As long as the labor and property are located in the United States, the supplier (that is, the workers and, for property, the owners) may be either U.S. residents or residents of foreign countries.

Heat content: The amount of heat energy available to be released by the transformation or use of a specified physical unit of an energy form (e.g., a ton of coal, a barrel of oil, a kilowatthour of electricity, a cubic foot of natural gas, or a pound of steam). The amount of heat energy is commonly expressed in British thermal units (Btu).

Heat pump: Heating and/or cooling equipment that, during the heating season, draws heat into a building from outside and, during the cooling season, ejects heat from the building to the outside. Heat pumps are vapor-compression refrigeration systems whose indoor/outdoor coils are used reversibly as condensers or evaporators, depending on the need for heating or cooling.

Heat rate: A measure of generating station thermal efficiency commonly stated as Btu per kilowatthour. Note: Heat rates can be expressed as either gross or net heat rates, depending whether the electricity output is gross or net generation. Heat rates are typically expressed as net heat rates.

Insulation: Any material or substance that provides a high resistance to the flow of heat from one surface to another.

The different types include blanket or batt, foam, or loose fill, which are used to reduce heat transfer by conduction. Dead air space is an insulating medium in storm windows and storms as it reduces passage of heat through conduction and convection. Reflective materials are used to reduce heat transfer by radiation.

Jacket: The enclosure on a water heater, furnace, or boiler.

Kinetic energy: Energy available as a result of motion that varies directly in proportion to an object's mass and the square of its velocity.

Langley: A unit or measure of solar radiation; 1 calorie per square centimeter or 3.69 Btu per square foot.

Lease fuel: Natural gas used in well, field, and lease operations, such as gas used in drilling operations, heaters, dehydrators, and field compressors.

Mains: A system of pipes for transporting gas within a distributing gas utility's retail service area to points of connection with consumer service pipes.

Mcf: One thousand cubic feet.

Mercaptan: An organic chemical compound that has a sulfur like odor that is added to natural gas before distribution to the consumer, to give it a distinct, unpleasant odor (smells like rotten eggs). This serves as a safety device by allowing it to be detected in the atmosphere, in cases where leaks occur.

Miscellaneous petroleum products: Includes all finished

products not classified elsewhere (e.g., petrolatum lube refining byproducts (aromatic extracts and tars), absorption oils, ram-jet fuel, petroleum rocket fuels, synthetic natural gas feedstocks, and specialty oils).

Moderator: A material, such as ordinary water, heavy water, or graphite, used in a reactor to slow down high-velocity neutrons, thus increasing the likelihood of further fission.

Naphtha: A generic term applied to a petroleum fraction with an approximate boiling range between 122 degrees Fahrenheit and 400 degrees Fahrenheit.

Octane: A flammable liquid hydrocarbon found in petroleum. Used as a standard to measure the anti-knock properties of motor fuel.

Ohm's Law: In a given electrical circuit, the amount of current in amperes is equal to the pressure in volts divided by the resistance, in ohms. The principle is named after the German scientist Georg Simon Ohm.

Ozone precursors: Chemical compounds, such as carbon monoxide, methane, nonmethane hydrocarbons, and nitrogen oxides, which in the presence of solar radiation react with other chemical compounds to form ozone.

Paraffin (oil): A light-colored, wax-free oil obtained by pressing paraffin distillate.

Paraffin (wax): The wax removed from paraffin distillates by chilling and pressing. When separating from solutions, it is a colorless, more or less translucent, crystalline mass, without odor and taste, slightly greasy to touch, and consisting of a mixture of solid hydrocarbons in which the paraffin series predominates.

Recycling: The process of converting materials that are no longer useful as designed or intended into a new product.

Refinery: An installation that manufactures finished petroleum products from crude oil, unfinished oils, natural gas liquids, other hydrocarbons, and oxygenates.

Reinjected: The forcing of gas under pressure into an oil reservoir in an attempt to increase recovery.

Renewable energy resources: Energy resources that are naturally replenishing but flow-limited. They are virtually inexhaustible in duration but limited in the amount of energy that is available per unit of time. Renewable energy resources include: biomass, hydro, geothermal, solar, wind, ocean thermal, wave action, and tidal action.

Repressuring: The injection of gas into oil or gas formations to effect greater ultimate recovery.

Reprocessing: Synonymous with chemical separations.

Roof (coal): The rock immediately above a coal seam. The roof is commonly a shale, often carbonaceous and softer than rocks higher up in the roof strata.

Spinning reserve: That reserve generating capacity running at a zero load and synchronized to the electric system.

Stack: A tall, vertical structure containing one or more flues used to discharge products of combustion to the atmosphere.

Surface mine: A coal-producing mine that is usually within a few hundred feet of the surface. Earth above or around the coal (overburden) is removed to expose the coalbed, which is then mined with surface excavation equipment, such as draglines, power shovels, bulldozers, loaders, and augers. It may also be known as an area, contour, open-pit, strip, or auger mine.

Tailings: The remaining portion of a metal-bearing ore consisting of finely ground rock and process liquid after some or all of the metal, such as uranium, has been extracted.

Thermal efficiency: A measure of the efficiency of converting a fuel to energy and useful work; useful work and energy output divided by higher heating value of input fuel times 100 (for percent).

Tie line: A transmission line connecting two or more power systems.

Ultraviolet: Electromagnetic radiation in the wavelength range of 4 to 400 nanometers.

Underground storage: The storage of natural gas in underground reservoirs at a different location from which it was produced.

Volatile solids: A solid material that is readily decomposable at relatively low temperatures.

Vacuum distillation: Distillation under reduced pressure (less the atmospheric) which lowers the boiling temperature of the liquid being distilled. This technique with its relatively low temperatures prevents cracking or decomposition of the charge stock.

Wafer: A thin sheet of semiconductor (photovoltaic material) made by cutting it from a single crystal or ingot.

Wattmeter: A device for measuring power consumption.

From The Energy Information Administration: Official Energy Statistics from the U.S. Government **(http://www. eia.doe.gov/glossary/index.html)**

BIBLIOGRAPHY

2.009 Archimedes Death Ray: Testing with MythBusters. Massachusetts Institute of Technology. **http://web. mit.edu/2.009/www//experiments/deathray/10_ Mythbusters.html**. 28 November 2007.

Archimedes' Weapon. TIME. **http://www.time. com/time/magazine/article/0,9171,908175,00. html?promoid=googlep**. 28 November 2007.

Back to the Future II. Universal Pictures, 1989.

Biodisel Basics. Union of Concerned Scientists. **http:// www.ucsusa.org/clean_vehicles/big_rig_cleanup/ biodiesel.html**. 29 November 2007.

Biogas Train. Svensk Biogas AB. **http://www. tekniskaverken.se/sb/docs/english/Biogastrain_ produktblad_2005.pdf**. (PDF version of document downloaded 19 November, 2007).

Biomass Program. U.S. Department of Energy: Energy Efficiency and Renewable Energy. **http://www1.eere. energy.gov/biomass/**. 1 December 2007.

Bonanno, A., and Schlattl, H., and Patern, L. *TheAge of the Sun and the Relativistic Corrections in the EOS.* **http:// www.aanda.org/index.php?option=article&access=st andard&Itemid=129&url=/articles/aa/full/2002/30/ aa2598/aa2598.right.html**. 28 November 2007.

Boyle, Godfrey, ed. *Renewable Energy: Power for a Sustainable Future.* Second Edition. New York: Oxford University Press, Inc., 2004.

Clinical Manifestations. Medical Microbiology. 4th Edition. **http://www.ncbi.nlm.nih.gov/books/bv.fcgi? rid=mmed.section.2222**. 1 December 2007.

Electricity From Biomass. Financial Times Management Report. 1998. p. 35.

Elliot, D.L., and Schwartz, M.N. "Wind Energy Potential in the United States." <u>PNL-SA-23109</u>. Sept. 1993

Geller, Howard. *Energy Revolution: Policies for a Sustainable Future.* Washington, Island Press, 2003.

Hurricane Research Division. Atlantic Oceanographic and Meteorological Library. **http://www.aoml.noaa.gov/hrd/ tcfaq/D7.html**. 3 December 2007.

Komor, Paul. *Renewable Energy Policy.* New York: Diebold Institute for Public Policy Studies, 2004.

Sørensen, Bent. *Renewable Energy: Its Physics, Engineering, Environmental Impacts, Economics & Planning.* Amsterdam: Elsevier Academic Press, 2004.

The Active Solar House. Construct Ireland. **http://www.**

constructireland.ie/articles/0213activesolarhouse. php. 22 November 2007.

The History of Solar. U.S. Department of Energy: Energy Efficiency and Renewable Energy. **http://www1.eere. energy.gov/solar/pdfs/solar_timeline.pdf**. (PDF version of document downloaded 21 November, 2007).

U.N. Environment Programme. **http://www.unep.org/tnt- unep/toolkit/Actions/Tool14/index.html**. 29 November 2007.

Energy Information Administration: Official Energy Statistics from the U.S. Government **http://www.eia.doe. gov/glossary/index.html**

AUTHOR BIOGRAPHY

David Craddock lives with his fiancé, Amie, in Daly City, California. Though David has had several articles published during his career as a journalist, *Renewable Energy Made Easy* is his first published book.

INDEX

DID YOU BORROW THIS COPY?

Have you been borrowing a copy of *Renewable Energy Made Easy: Free Energy from Solar, Wind, Hydropower, and Other Alternative Energy Sources* from a friend, colleague, or library? Wouldn't you like your own copy for quick and easy reference? To order, photocopy the form below and send to:

Atlantic Publishing Company
1405 SW 6th Ave • Ocala, FL 34471-0640